GOVERNANCE AND CONTROL
OF FINANCIAL SYSTEMS

Ashgate Studies in Resilience Engineering

Series Editors

Professor Erik Hollnagel, *Institute of Public Health,
University of Southern Denmark, Denmark*

Sidney Dekker, *Professor and Director, Key Centre for Ethics, Law, Justice
and Governance, Griffith University, Brisbane, Australia*

Christopher P. Nemeth, *PhD, CHFP, Principal Scientist and Group Leader,
Cognitive Systems Engineering, Klein Associates Division (KAD)
of Applied Research Associates (ARA), Dayton, Ohio, USA*

Dr Yushi Fujita, *Technova, Inc., Japan*

Resilience engineering has become a recognized alternative to traditional approaches to safety management. Whereas these have focused on risks and failures as the result of a degradation of normal performance, resilience engineering sees failures and successes as two sides of the same coin – as different outcomes of how people and organizations cope with a complex, underspecified and therefore partly unpredictable environment.

Normal performance requires people and organizations at all times to adjust their activities to meet the current conditions of the workplace, by trading-off efficiency and thoroughness and by making sacrificing decisions. But because information, resources and time are always finite such adjustments will be approximate and consequently performance is variable. Under normal conditions this is of little consequence, but every now and then – and sometimes with a disturbing regularity – the performance variability may combine in unexpected ways and give rise to unwanted outcomes.

The Ashgate Studies in Resilience Engineering series promulgates new methods, principles and experiences that can complement established safety management approaches. It provides invaluable insights and guidance for practitioners and researchers alike in all safety-critical domains. While the Studies pertain to all complex systems they are of particular interest to high-hazard sectors such as aviation, ground transportation, the military, energy production and distribution, and healthcare.

Governance and Control
of Financial Systems
A Resilience Engineering Perspective

Edited by

GUNILLA SUNDSTRÖM
Deutsche Bank, Germany

ERIK HOLLNAGEL
Institute of Public Health, University of Southern Denmark, Denmark

CRC Press
Taylor & Francis Group
Boca Raton London New York

CRC Press is an imprint of the
Taylor & Francis Group, an **informa** business

CRC Press
Taylor & Francis Group
6000 Broken Sound Parkway NW, Suite 300
Boca Raton, FL 33487-2742

First issued in paperback 2017

© 2011 by Gunilla Sundström and Erik Hollnagel
CRC Press is an imprint of Taylor & Francis Group, an Informa business

No claim to original U.S. Government works

Version Date: 20160226

ISBN 13: 978-1-4094-2966-1 (hbk)
ISBN 13: 978-1-138-07448-4 (pbk)

Visit the Taylor & Francis Web site at
http://www.taylorandfrancis.com

and the CRC Press Web site at
http://www.crcpress.com

Contents

List of Figures and Tables   *vii*
List of Contributors   *ix*
Prologue   *xiii*

Chapter 1   The Turmoil in the Financial Services System   1
                 Gunilla Sundström and Erik Hollnagel

SECTION I: **UNDERSTANDING WHY:**
                 THE NEED FOR NEW PERSPECTIVES   7
                 Gunilla Sundström and Erik Hollnagel

Chapter 2   What is the Financial Services System?   11
                 Gunilla Sundström and Erik Hollnagel

Chapter 3   A Dynamic Systems Modelling Perspective   17
                 Gunilla Sundström and Erik Hollnagel

Chapter 4   From the Efficient Market Hypothesis to Econophysics 27
                 Bill McKelvey and Rossitsa Yalamova

SECTION II: **UNDERSTANDING WHAT:**
                 MAKING SENSE OF UNPREDICTABLE
                 EVENTS AND DEVELOPMENTS   37
                 Erik Hollnagel and Gunilla Sundström

Chapter 5   The 2007 Liquidity Crisis: An Example of Scalability
                 Dynamics in Action   41
                 Bill McKelvey and Rossitsa Yalamova

Chapter 6   Taming Manias: On the Origins, Inevitability,
                 Prediction and Regulation of Bubbles and Crashes   55
                 Jeff Satinover and Didier Sornette

Chapter 7   Using Power Laws and the Hurst Exponent
to Identify Stock Market Trading Bubbles   85
Rossitsa Yalamova and Bill McKelvey

**SECTION III: UNDERSTANDING HOW:
TURNING FINANCIAL SERVICES
SYSTEMS INTO RESILIENT SYSTEMS**   **107**
Erik Hollnagel and Gunilla Sundström

Chapter 8   Balancing Different Modes of Uncertainty
Management in the Financial Services Industry   111
Gudela Grote

Chapter 9   Financial Resilience Engineering: Toward Automatic
Action Formulas against Risk and Reckless
Endangerment   133
Bill McKelvey and Rossitsa Yalamova

Chapter 10   The Ability to Regulate, Govern and Control
Financial Services Systems   149
Gunilla Sundström and Erik Hollnagel

Epilogue: Financial Markets and the Law of Requisite Variety   *165*
References   *169*
Index   *187*

List of Figures and Tables

Figures

1.1   'Phases' of the 2007–2009 Financial Services system crisis   4

3.1   Control of the economy by negative feedback   20
3.2   Mutually coupled processes   22
3.3   The impact of changes in the confidence of valuations   24
3.4   The destructive power of pro-cyclicality   25

Section II.1   Zöllner's illusion   38

6.1   Variation in volatility in the MG   65
6.2   Illustration of the concept of Dragon Kings in the
       financial sector   68
6.3   Rank-ordering plot of the population of French cities   70
6.4   Typical set of optimal configurations for optimal paths
       of length W=4,096 and for 0≤y≤1,200   72
6.5   Schematic representation of 'avalanches' between
       successive optimal RDP paths fixed at their two end points   74
6.6   Probability density function P(S) of avalanche sizes S
       as a function of the rescaled variable S/W5/3 for lattice
       width W varying from 8 to 512 on a log-log plot   75
6.7   Draw-downs (+) in the NASDAQ Composite index   78
6.8   Ensemble of most-likely LPPL fits for a collapse at the
       80 per cent confidence level   79

7.1   Financial markets phase diagram   87
7.2   Depiction of volatility incidents above GARCH line   94
7.3   Power-law fit DJIA (# events vs. per cent daily change)   96
7.4   DJIA daily returns 1928–2007   98
7.5   Log-periodicity in the Hang-Seng stock market
       (1970–2000)   103

8.1   Basic principles of uncertainty management   115

10.1   Three generic system states identified by state variables   152
10.2   An example of a FRAM-based functional representation   160

Tables

1.1   Scope and impact of 2007–2009 financial crisis   6

5.1   The four feedback loops   44
5.2   SFTs associated with the crisis build-up   47

List of Contributors

Gudela Grote is Professor of Work and Organisational Psychology in the Department of Management, Technology, and Economics at the ETH Zurich. She holds a Master's degree in psychology from the Technical University in Berlin and a PhD in Industrial/Organisational Psychology from the Georgia Institute of Technology, Atlanta. She has published widely on the interplay of organisation and technology, safety management, and changing employment relationships. Gudela Grote is Associate Editor of the journal *Safety Science*. Special interests in her research are the increasing flexibility and virtuality of work and their consequences for the individual and organisational management of uncertainty.

Erik Hollnagel is Professor at the University of Southern Denmark, Professor and Industrial Safety Chair at MINES ParisTech (France), and Visiting Professor at the Norwegian University of Science and Technology (NTNU) in Trondheim (Norway). He has worked at universities, research centres, and industries in several countries and with problems from many domains. His professional interests include industrial safety, resilience engineering, accident investigation, cognitive systems engineering and cognitive ergonomics. He has published widely and is the author/editor of 18 books, including 4 books on resilience engineering. The latest title from Ashgate is *Resilience Engineering in Practice: A Guidebook*.

Bill McKelvey, PhD MIT 1967. Professor of Strategic Organizing and Complexity Science at the UCLA Anderson School of Management. His book, *Organisational Systematics* (1982) remains the definitive treatment of organisational taxonomy and evolutionary theory. He chaired the building committee that produced the $110,000,000 Anderson Complex at UCLA – opened in 1995. In 1997 he became

Director of the Center for Rescuing Strategy and Organization Science (SOS). From this Center he initiated activities leading to the founding of UCLA's Inter-Departmental Program, Human Complex Systems and Computational Social Science. He has directed over 170 field study teams on six-month projects concerned with strategic and organisational improvements to client firms. Co-edited *Variations in Organisation Science* (with J. Baum, 1999), a special issue of *Emergence* (with S. Maguire, 1999), and a special issue of *J. Information Technology* (with J. Merali, 2006). Co-editor of *SAGE Handbook of Complexity and Management* (2010); Editor of *Routledge Major Work: Complexity* (2011; 5-volumes, 2,000 pp.). Articles appear in: *Admin. Sci. Quart.; Organ. Science; Acad. Mgmt. Rev.; J. Bioeconomics; Leadership Quart.; J. Behavioral Finance, Academic Questions; J. Management; Strategic Organization; Nonlinear Dynamics; Psych. and Life Sci.; J. Int. Business Studies; Int. J. Production Economics; J. Information Technology; Research in Competence-Based Mgmt.; J. Economics and Org. Behavior; Emergence; Int. J. Accounting and Info. Mgmt.; Advances in Strategic Mgmt.; J. Business Venturing; Emergence: Complexity and Organization; Proceed. Nat. Acad. of Sciences; Risk Management, An Int. J.;* among others. He has approximately 70 recent papers applying complexity science to organisation science and management.

Jeffrey Satinover, MD, PhD, is Distinguished Professor of Mathematics and Science at the King's College in New York and visiting scientist at the Swiss Federal Technical Institute and the Lorange School of Business in Zurich, Switzerland. He is Managing Director of Quintium Analytics, Ltd. Dr Satinover specialises in complex systems theory and in the design of new risk management methods. He holds degrees from MIT, Harvard, Yale and the University of Nice and has taught at Harvard, Yale and Princeton.

Didier Sornette is Professor of Entrepreneurial Risks in the Department of Management, Technology and Economics at the Swiss Federal Institute of Technology (ETH Zurich), a professor of finance at the Swiss Finance Institute, a professor of Physics and a professor of Geophysics also at ETH Zurich. He is the author 450+ research papers and 5 books. His research focuses on the prediction of crises and extreme events in complex systems and in particular of financial bubbles and crashes, and the diagnostic of systemic instabilities. Other applications include earthquake physics and geophysics, financial economics and the theory of complex systems, the dynamics of success on social networks and the complex system approach to medicine (immune system, epilepsy). He recently launched the Financial Crisis Observatory to test the hypothesis that financial bubbles can be diagnosed in real time and their termination can be predicted probabilistically.

Gunilla A. Sundström is a global leader with a passion for Financial Services, Outsourcing/Offshoring and R&D. She has a leadership, execution and innovation track record in several areas including outsourcing/offshoring, decision support and analytics for operations, resilience engineering, governance and risk management. She has held leadership positions in a variety of industries including R&D, Financial Services and Telecommunications. She currently holds the position as the Smartsourcing and Technology Services Leader at Deutsche Bank. Gunilla Sundström has published more than 60 papers; holds two US Patents and has been awarded IEEE-Systems, Man and Cybernetics' outstanding contributions award. She currently serves on the editorial boards of the *International Journal of Human Computer Interaction* and the *International Journal of Cognition, Technology and Work*. She holds a Dr Phil degree from University of Mannheim, Germany.

Rossitsa M. Yalamova is an assistant professor of Finance at the University of Lethbridge in Alberta, Canada. She holds a PhD in finance from Kent State University and MD from Saint Petersburg State Medical Academy, Russia. Her research has been published in *Fractals, Investment Management and Financial Innovations, International Research Journal of Finance and Economics,* and the *Asian Academy of Management Journal of Accounting and Finance*. She is interested in market crashes, non-linear dynamics, complex networks and chaos. She was a participant in the 2007 Santa Fe Institute Complex Systems Summer School held at the Beijing Institute of Theoretical Physics and a visiting professor in the Facoltà di Scienze Economiche, Università della Svizzera Italiana, Lugano, Switzerland in 2009/10.

Also available:

Resilience Engineering Perspectives:
Volume 1: Remaining Sensitive to the Possibility of Failure
Edited by Erik Hollnagel, Christopher P. Nemeth
and Sidney Dekker
ISBN: 978-0-7546-7127-5 (hbk)

Resilience Engineering Perspectives:
Volume 2: Preparation and Restoration
Edited by Christopher P. Nemeth, Erik Hollnagel
and Sidney Dekker
ISBN: 978-0-7546-7520-4 (hbk)

Resilience Engineering in Practice:
A Guidebook
Edited by Erik Hollnagel, Jean Pariès, David D. Woods
and John Wreathall
ISBN: 978-1-4094-1035-5 (hbk)
ISBN: 978-1-4094-1036-2 (ebk)

Resilience Engineering:
Concepts and Precepts
Edited by Erik Hollnagel, David D. Woods and Nancy Leveson
ISBN: 978-0-7546-4641-9 (hbk)
ISBN: 978-0-7546-4904-5 (pbk)
ISBN: 978-0-7546-8136-6 (ebk)

Prologue

Gunilla Sundström and Erik Hollnagel

In late 2006 and early 2007, two of largest providers of subprime lending products in the USA (that is, New Century Financial Corporation and Countrywide Financial) had to declare bankruptcy. In April 2007, two hedge funds managed by Bear Stearns, a US Investment Bank, collapsed. In September 2007, the UK-based financial services firm Northern Rock had to request emergency funds from the Bank of England. All these events seemed to have a common denominator, namely the US subprime lending market. This was a lending market that focused on providing funds to so-called subprime borrowers with economic limited resources. Typical products offered on these markets were home loans with adjustable interest rates, the underwriting procedures used were less stringent, and borrowers were more risky. Subprime market participants commonly made the assumption that house prices would continue to appreciate and that as a result the subprime market would continue to thrive. However, the US subprime market collapsed in 2007, an event that triggered previously unimaginable conditions and events in the global Financial Services system and the world economy. As this book is published, the impact of these events is still felt across the globe.

Central Banks, the Financial Stability Board, the G20, researchers and many others quickly tried to provide explanations about 'what happened' and what needs to be done to prevent something similar from happening in the future. This seemed to follow the astute comment by Nietzsche (1895) that when people face the unknown they enter a state of danger, disquiet, and anxiety, where:

the first instinct is to eliminate these distressing states. First principle: any
explanation is better than none ...

To illustrate that, the Financial Stability Board published a report
in April 2009 stating that 'procyclicity' played a major role,
that is, the mutually reinforcing feedback mechanisms between
the Financial Services system (FSS) and the global economy
(www.financialstabilityboard.org). In their book, Acharaya and
Richardson (2009: 2) argued that 'subprime defaults were the root
cause, the most identifiable event that led to systemic failure was
most likely the collapse on June 2007 of two Bear Stearns managed
hedge funds'. In the so-called de Larosière report published by
the European Commission in February 2009 (de Larosière, 2009),
a section of the report is devoted to 'Causes of the Financial
Crisis'. Rather than singling out specific events, a list of parallel
developments is outlined and the key point is made that the
crisis developed in a dynamic way by the combination of many
conditions including excess capital, low interest rates, financial
innovations and excessive growth of global credit markets.

The natural first questions after such a crisis of the global
Financial Services system were what actually happened and
why? These two questions were then of course followed by the
question of whether it could happen again. One purpose of the
present book is to provide some answers to these three questions.
A second, and equally important, purpose is to put the questions
themselves under scrutiny. This will be done by bringing together
concepts from a variety of disciplines and areas, including
industrial safety, in order to answer three questions:

- Why did the global Financial Services system experience
  the 2007–2009 turbulence? And did anyone know that this
  would happen?
- How can we know what is likely to happen next in the
  global Financial Services system? How do we know what is
  affected by what?
- How can we re-establish a sustainable performance of the
  global Financial Services system and improve its resilience?

Concepts and theories from resilience engineering play a key
role throughout the book (cf. Hollnagel, Woods and Leveson,

2006; Hollnagel et al., 2011). The field of resilience engineering emerged in part as a result of observations that success and failure are like 'yin and yang,' that is, success and failure should be understood simply as different manifestations of the same underlying events or processes. In other words, what makes us fail can also make us successful, because at the end of the day, positive and negative outcomes are the result of combinations of a virtually innumerable number of possible conditions or states. The difference is that highly desirable outcomes are seen as a result of goal-oriented behaviour while highly undesirable outcomes are perceived as the consequences of unexpected and even unimaginable events and conditions, rather than as a result of deliberate action plans.

Concepts from systems theory (von Bertalanffy, 1975) also play a major role in the present book. The goal of General Systems Theory was to identify the laws that govern the dynamics and behaviour of 'organised entities' such as organisations, social groups and technological devices. The concept of socio-technical system (for example, Trist, 1981) was developed using key assumptions from General Systems Theory. The concept of socio-technical systems is important in the present context since it explicitly recognises that system behaviour depends on the tight couplings among humans, organisations and technologies. Examples of such complex systems include aviation, energy power generation and distribution, industrial production, telecommunication, healthcare systems – and of course Financial Services systems. To perform effectively, a Financial Services system must constantly organise and re-organise its resources, people, processes and technologies around its objectives. In each organisational form people, processes and technology will be interconnected, sometimes because it was intended but more often because unexpected connections or couplings will emerge. Each stakeholder in the Financial Services system plays a role in how these connections emerge and, most importantly, stakeholders need to pay constant attention to the possible implications of the interconnectedness. The Financial Services system is, for better or worse, a Large-Scale Complex Socio-Technical System (LSCSTS).

The introductory chapter of the book provides a summary of what happened during the height of the 2007–2009 Financial

Services system crisis. The focus is on describing conditions of the global Financial Services system during various time periods as well as the events that recently occurred. Section I builds on Chapter 1 and provides some arguments why new perspectives are required to better understand what happened, what could happen, and how to make the Financial Services system more resilient. Section II is focused on how to make sense of seemingly unpredictable events, that is, to better understand what happened. In this section, McKelvey and Yalamova describe how many tiny initiating events, including technological innovation, eventually led to the 2007 liquidity crisis. Satinover and Sornette outline how mutually reinforcing 'bubbles' created a perception that outcomes were positive while they actually were not. Instead, outcomes were the result of an 'imaginary' machine that seemed to create wealth for everyone, but obviously instead led to unimaginable events such as the freezing of the global credit markets. In Section II's last chapter, Yalamova and McKelvey describe a method for how to detect bubble-build-ups in stock markets before they happen and thus open the door for proactive action plans to either prevent or dampen the impacts of bubbles. Section III is devoted to understanding how Financial Services systems can be made more resilient, that is, able to sustain their performance over time despite unexpected conditions and events. In this section, Grote outlines how a resilient system needs to be able to have more than one way to manage (or respond) to uncertainty; McKelvey and Yalamova describe so-called resilience engineering interventions, that is, actions that can be taken to either prevent bubbles from developing and/or actions that will dampen the impact of bubbles that have burst. Finally, Sundström and Hollnagel describe how we can learn from the study of dynamic systems and ways to control these leveraging feedback systems. The book concludes with a summary of key points made in the previous chapters and places them in the broader context of a resilience engineering perspective of Financial Services systems.

Chapter 1
The Turmoil in the Financial Services System

Gunilla Sundström and Erik Hollnagel

At the time of finishing this book, that is, December 2010, the impact of the 2007–2009 turmoil in the global Financial Services system (FSS) is still felt by the global economy and the Financial Services industry. In response to the crises, the industry continues to transform itself in front of our very eyes, trying to cope with something that has not yet been completely understood. National governments have intervened in various ways, referred to as rescue, bailout or economic stimulus policies and programmes, often reflecting political positioning as much as sound reasoning. Regulatory bodies such as the US Federal Reserve Bank, the European Central Bank, the Bank of Japan, the People's Bank of China, and the Bank of England have taken unprecedented measures to stabilise the global Financial Services system, to the best of their understanding. New regulatory bodies have emerged, such as the European Systemic Risk Board and the US-based Systemic Risk Regulatory Council, both chartered with monitoring of systemic risk; in addition, scope of existing regulatory bodies were broadened. All of these actions were of course taken in response to the financial crisis. While these efforts have been credited with preventing a complete meltdown of the global Financial Services system, it is not necessarily clear that demands for more regulation will lead to better control of the Financial Services system. In fact, some argue that measures taken by governments and regulatory bodies have prolonged the financial crisis and its impacts (for example, Taylor, 2009). So how can we know, or find out what actually happened? How do we know that we have improved our abilities to govern and control Financial Services systems?

What Happened?

One of the fundamental facts of perception and understanding, and indeed of epistemology, is that the perspective (or model) that you use to describe events or to understand why something happened, will influence the result of the analysis. This is particularly important in relation to the investigation of things that have gone wrong, because it means that accident investigation is a psychological rather than a logical exercise. Causes are constructed rather than found, and the financial crisis is no exception. Hollnagel and Speziali (2008) described this as the *What-You-Look-For-Is-What-You-Find* or the *WYLFIWYF* principle. The key implication of the *What-You-Look-For-Is-What-You-Find* in relation to the 2007–2009 financial crisis is that it is important to be aware of the assumptions that various people, or communities, use to describe the events. To demonstrate the importance of underlying assumptions, we will in the following show the consequences of adopting a so-called linear view.

The Linear View

The linear view assumes that it is possible to define a sequence of events such that events in the beginning of the sequence invariably lead to events later in the sequence. This has been expressed clearly as the First Axiom of Industrial Safety (Heinrich, 1959: 13), which reads:

> The occurrence of an injury invariably results from a completed sequence of factors – the last one of these being the accident itself.

The traditional root cause perspective is an example of a linear view in which events in the beginning of the sequence are viewed as causing events later in the sequence. It is consistent with this view that the collapse of the US subprime market was seen as a root cause of the financial crisis (for example, Acharaya and Richardson, 2009).

The US subprime market was part of the broader US mortgage market. As described by, for example, Jaffe (2008), the market experienced a tremendous growth in the years 2001–2007, that is, the portion of subprime loans reached approximately 20 per cent of all originated mortgages (see Jaffe, 2008). In addition to growth, the US subprime market also changed in that the proportion of originated subprime mortgages doubled from 2003 to 2006. In the US, a conventional loan typically had a fixed interest rate and ran for 30 years and the borrower made a down payment of about a 20 per cent of the price of the house. In contrast to that, a subprime mortgage often had an adjustable interest rate and no requirements for the borrower to provide a cash down payment. As a result, the default risks associated with the subprime market were considerably higher.

As already mentioned, the portion of risky loans grew drastically in the years preceding the financial crisis. However, lending organisations typically did not keep loans on their books, but sold the loans to financial intermediaries who in their turn pooled mortgages to offer products such as mortgage-backed securities (MBSs), collateralised mortgage obligations (CMOs) and collateralised debt obligations (CDOs). The result was that investors across the globe de facto became *connected* to the US subprime markets by investing in products that depended on the behaviour of that market, even though they might not have intended to do so or been aware of it.

The rapid expansion of the US mortgage market was fuelled by the assumption that house valuations would continue to appreciate and that borrowers therefore at some point could replace risky loans, such as adjustable rate loans, with less risky loans, such as fixed-interest-rate loans. In 2007 default rates started to increase in the US subprime mortgage market. In a linear cause-effect explanation of what happened, this event is widely perceived as one of the key triggers of the financial crisis (see Figure 1.1), which provides a simple linear view of some of the key phases at the height of the 2007–2009 financial crisis. The increased default rates undermined valuations and investors' confidence in mortgage-backed structured products dwindled. As a result financial markets for financial products exposed, or with assumed exposure, to the US subprime markets froze. Soon, investors' risk-aversion and

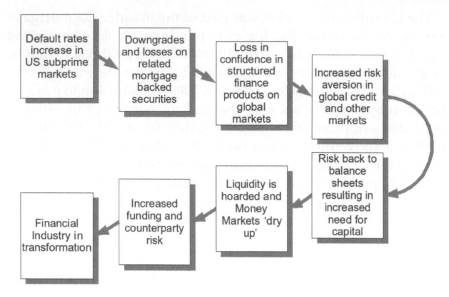

Figure 1.1   **'Phases' of the 2007–2009 Financial Services system crisis**

Source: Based on Bank of England's 2007 Financial Stability Report's chart 1.

weariness spilled over to other markets such as the short-term global credit markets. As a result major Financial Services firms faced increased liquidity risk, that is, the risk of not being able to turn assets into cash, or cash equivalents became higher. In parallel, firms' capital needs increased to offset risk associated with other types of assets on their balance sheets. Cash hoarding, risk-aversion and lost trust became the characteristics of many key global Financial Services firms. As a result financial markets froze for financial products and firms that were exposed, or were assumed to be exposed, to the US subprime markets.

These industry-wide tensions were, for example, very visible in the high inter-banking interest rates (that is, LIBOR). In parallel, complex asset classes experienced continued devaluation leading to an even greater need for capital. As a result, some firms started to shed assets. An example is that the US-based Financial Services Company Merrill Lynch (later acquired by Bank of America) sold $30 billion worth of assets in July–August 2008. Such fire-sales further accelerated asset devaluations, and led to yet more need for capital.

Central banks reacted to the crisis by cutting interest rates and by putting different vehicles in place to improve liquidity. National governments implemented stimulus, bailout, and rescue packages, all designed to prevent the global Financial Services system from a complete collapse. Table 1.1 provides a view of the scope of the crisis by describing how a selected set of Financial Services firms were impacted. Three types of events are distinguished, all of them associated with different levels of distress experienced by a Financial Services firm:

1. Acquisition, that is, one Financial Services firm acquires another firm.
2. Bankruptcy filing/nationalisation, that is, a firm filed for bankruptcy, or was nationalised by the government.
3. Bailout, that is, a firm leveraged government provided capital.

Table 1.1 represents a high-level view of selected key events in the global Financial Services system from the beginning of 2007; however, the event list is not exhaustive. For example, it does not include the smaller US banks closed by the US regulatory body FDIC (Federal Deposit Insurance Corporation) in 2009, that is, 141; or the more than 157 banks closed by December 2010 in the US by FDIC. Table 1.1 nevertheless provides a good view of both the number and the types of firms impacted in some of the major developed Western economies.

Looking at the list, the obvious question is: why did the global Financial Services system experience the 2007–2009 turbulence? In Section I of this book, different perspectives are leveraged to provide insights into why the global Financial Services system drifted towards complete systemic collapse.

Although linear explanations implying sequences or chains of causes and effects were introduced as explanations from the very beginning of the crisis, even the brief exposure here makes it clear that they are too simple to be reasonable. If the Financial Services system worked as the clockwork implied by the simple linear models, then it would have been possible effectively to intervene and to change the course of events. As Chapters 3 and 4 will show, there is a need for much more powerful explanations.

Section II will present some of these in more detail, while Section III will return to the fundamental issue of governance and control of Financial Services systems and point to several ways forward.

Table 1.1   Scope and impact of 2007–2009 financial crisis

	2007	2008	2009	2010
Bankrupt, nationalised	Northern Rock (UK) Victoria Mortgage Funding (UK) New Century Financial Corp (US) American Home Mortgage (US)	Lehman Brothers (US) Washington Mutual (US) Northern Rock (UK) IndyMac (US) Fannie Mae/ Freddie Mac (US) Glitnir Bank (ICE) Landebanki (ICE) Bradford and Bingley (UK) Fortis (B, NL, LUX) ABN Amro – Dutch assets (NL)	Anglo Irish Bank (IRL) Hypo Real Estate (D) Hypo Group Alpe Adira (A)	
Government bailout	Landesbank Sachsen (D)	AIG (US) Citigroup (US) Wells Fargo (US) GMAC (US) AMEX (US) Bank of America (US) RBS (UK) HBOS (UK) Lloyds Banking Group (UK) Morgan Stanley (US)	Caja Castilla La Mancha (E)	
Acquired	ABN AMRO (NL)	Bear Stearns (US) Merill Lynch (US) Wachovia (US) National City (US) Countrywide (US) Washington Mutual (US) First Charter Bank (US)	Barclays Global Invest (UK)	ABN Amro (partial acquisition) (NL) Fortis Bank/ABN Amro (NL)

Sources: Bank of England (2009) and Federal Reserve Bank St Louis (2009).

SECTION I
Understanding Why:
The Need for New Perspectives

Introduction

Gunilla Sundström and Erik Hollnagel

In order to understand what happened in the recent financial crisis we must go beyond a simple sequence of events aligned on a timeline. What we need is a framework that helps us to understand the behaviour of Financial Services systems as complex socio-technical systems, and most important of all – a framework that enables us to improve the resilience of the Financial Services system and thereby make them better able to adjust to and continue functioning when unexpected conditions arise – as they are certain to do.

The previously introduced *What-You-Look-For-Is-What-You-Find* (WYLFIWYF) principle states the fundamental fact that what we expect to see more or less determines what we find. Although seemingly trivial, the principle nevertheless provides us with the critical awareness that the assumptions we make define how we characterise individual events and find patterns in how events develop over time. In the Financial World several key assumptions have dominated the thinking about Financial Markets for most of the past decades. Two of the most important ones are the assumption about efficient markets (Fama, 1970) and the assumption about the independence of events. The first assumption is the basis of the belief that markets will behave in a rational manner (that is, *homo economicus*), leveraging available information and processing it in an efficient and rational way so that it remains in a healthy state. The second assumption supports the belief that risks associated with individual asset classes, even if they are very similar, are independent of each other, that is, that the risk of owing an asset from one class can be offset by owing an asset from a similar class. The principle of hedging as a way to mitigate risk is based on the assumption that you can offset the risk of owning one asset class by also owning assets that you expect to move in the opposite direction.

Adopting a resilience engineering perspective enables us to look at Financial Services systems in a different way, and therefore see different things. The field of resilience engineering is based on the four following premises:

- Performance conditions are always underspecified. Individuals and organisations must therefore always adjust their performance to match current demands and resources; because resources and time are finite, such adjustments will inevitably be approximate.
- Many adverse events can be attributed to a breakdown or malfunctioning of components and normal system functions, that is, to an identifiable cause, but many cannot. These events should instead be understood as the outcome of unexpected combinations of performance variability.
- Safety management, or management of uncertainty, cannot be based on hindsight, nor rely on error tabulation and the calculation of failure probabilities. Safety management must be proactive as well as reactive.
- Safety cannot be isolated from the core (business) process, nor vice versa. Safety is the prerequisite for productivity, and productivity is the prerequisite for safety. Safety is achieved by improvements and empowerment rather than by constraints.

The field of resilience engineering was developed to deal with problems of safety in industrial systems. But the reasoning that applies to industrial safety can a fortiori be applied to the safety of Financial Services systems. If we therefore try to look at the Financial Services system with these premises in mind, we can make the following observation:

- Systems vary with respect to their coupling and tractability. The meaning of coupling is that subsystems and/or components are connected or depend upon each other in a functional sense (Perrow, 1984). Tightly coupled systems are difficult to control because an event in one part of the system quickly will spread to other parts. A tightly coupled system is therefore more likely to experience a rapid propagation of the impact of outcomes than a loosely coupled system. The way that the impact of the collapse of the US subprime market affected the entire Financial Services system and led to unimaginable consequences for companies, individuals and economies, is an illustration of how a set of outcomes 'travel' through a tightly coupled system.

Tractability means that a system is easy to comprehend and therefore easy to govern and/or manage (Hollnagel, 2009a). The Financial Services system is clearly not tractable and is unlikely to ever be tractable. On the contrary, descriptions of Financial Services systems are elaborate with (too) many details, the principles of functioning are not completely known, and the dynamics of the Financial Services systems are such that they change faster than they can be described – the efficient market assumption notwithstanding. The Financial Services system is underspecified and therefore intractable – difficult to comprehend and difficult to govern or manage. (Attempts to provide oversimplified explanations such as the one shown in Figure 1.2 do not make the system tractable. The explanations are simply wrong because they are oversimplified.)

In Chapter 2 we discuss two commonly used perspectives, the institutional and the functional, to describe Financial Services systems and suggest that General Systems Theory provides helpful concepts to analyse Financial Services systems:

- Since Financial Services systems are complex socio-technical systems they must be viewed as dynamic open systems and should be analysed and modelled as any other dynamic open system. Chapter 3 describes how simple modelling principles can be leveraged to create a deeper understanding of the dynamics associated with Financial Services systems.

A better understanding of the possible dynamics will not only improve our understanding of what happened during the financial crisis but also help us generate ideas for how to make the Financial Services system more resilient.

- The ability of a system to proactively understand the impact of its behaviours is critical from a resilience engineering perspective. (It corresponds to the ability to anticipate; see Hollnagel, 2009b). In Chapter 4 concepts from the world of econophysics are introduced to illustrate the search for ways to proactively detect financial bubbles.

The three chapters in Section I provide an understanding of why the turmoil in the Financial Services system occurred. Section II continues from this basis by looking at how we can make sense of unpredictable events and developments, not by explaining them in terms of simple cause-effect relations, but by understanding the dynamics of complex socio-technical systems. Finally, Section III completes this line of reasoning by explaining how we can improve the resilience of Financial Services systems.

Chapter 2
What is the Financial Services System?

Gunilla Sundström and Erik Hollnagel

The suggestion that a *shadow* banking system contributed to the 2007–2009 crisis in the Financial Services system only makes sense if it is possible to describe the alternative, that is, the *real* Financial Services system. This view, often referred to as an institutional perspective (for example, Merton, 1995), is based on the belief that it is possible to define a set of attributes, and use these to decide if a particular organisation is a part of the Financial Services system or not. The institutional perspective 'takes as given the existing institutional structure of financial intermediaries and views the objective of public policy as helping the institutions currently in place to survive and flourish' (Merton, 1995: 23). The institutional perspective does not require very elaborate descriptions, since the principles of functioning are assumed to be known. And since the institutional structure is stable, the institutional perspective implies that the Financial Services system is tractable (see Introduction to Section I).

The challenges associated with the institutional perspective, in a sense the challenges from making the assumption that Financial Services systems are tractable, are to define the precise attributes and to recognise all organisations that posses these attributes. The events during the height of the 2007–2009 crisis clearly demonstrated that neither of these challenges could be met. This need not have been a surprise, since a little logical reasoning also leads to the same conclusion, namely that the institutional perspective does not provide a good foundation for a definition of the Financial Services system. In his popular book published in 2007, Taleb introduced the *Black Swan logic* to argue that 'Black Swan logic makes *what you don't know* far more relevant than what you do know' (Taleb, 2007: XIX). The Black Swan logic refers to the classical philosophical problem known as the Problem of Induction (for example, Popper, 1967). Simplified, the Problem of Induction is whether inductive reasoning can lead to knowledge, which more practically means when we can reasonably infer that a generalisation is valid. (The answer, of course, is that we can never do that.) For example, how do we know that the generalisation 'All swans are white' is a valid or true statement? Is it sufficient to see 10 white swans in one location? Or do we need to see 100 or 1,000 swans? May we conclude that all swans are white, because all swans that we have ever seen have been white? Obviously neither situation provides us with the necessary logical justification to infer that all swans are white. For example, as we watch a flock of white swans, a black swan might suddenly join the group. Or, while unlikely, the swan we see tomorrow could be black. In both cases, the generalisation that 'All swans are white' has to be rejected. The key point here is that while inductive reasoning may support a conclusion, it can never prove that it is true. All it can do is to reinforce our belief that the generalisation is true. Unfortunately this brings into play the so-called confirmation bias, which is the name for the powerful psychological tendency to look for information that is consistent with an existing belief (cf. Klayman and Ha, 1987). This is, more generally, also an example of the efficiency-thoroughness trade-off (Hollnagel, 2009a), in the sense that it requires less time and effort to look for confirming evidence (= efficiency) than to look for evidence that may contradict an existing belief (= thoroughness). For example, if we observe a flock of white swans we could either look for one more white swan to confirm our belief, or we could try to look – far and wide – for swans of a different colour. In the case of Financial Services systems, the belief is that all Financial Services systems have the qualities defined by the institutional perspective, hence that there is something that we can contrast to the 'shadow banking system'.

   Merton (1995) introduced the functional perspective as an alternative to the institutional view of Financial Services systems. 'The functional perspective takes as given the economic functions performed by financial intermediaries and asks what is the best institutional structure to perform those functions' (Merton, 1995). By focusing on the functions rather than the structures, Financial Services systems are no longer defined in terms of a number of attributes, that is, as financial institutions. The functional perspective proposes:

1. that financial functions are more stable than financial institutions, hence change less over time, and
2. that financial functions change to accommodate the effects of external forces, notably competition.

The emphasis on functions rather than structures is similar to one of the main assumptions of resilience engineering that resilience is not a system property but that it characterises what a system *does* rather than what it has (Hollnagel, Woods and Leveson, 2006). The functional perspective also means that Financial Services systems under conditions of rapid change, such as during the financial crisis, become intractable.

By adopting this perspective, the focus shifts from the particular attributes of the organisation performing the function to the functions performed. The type of functions performed are all related to the goal of a Financial Services system, which is 'to facilitate the allocation and deployment of economic resources, both spatially and temporally, in an uncertain environment' (Merton, 1995: 23). In other words, any function that allocates and/or supports deployment of economic resources is a Financial Services function. How the allocation and/or deployment of economic resources occurs, or who performs the reallocation and/or the deployment, is not the focus. The more precise set of functions that are required to facilitate allocation and deployment of economic resources depends on the particular model decision-makers and stakeholders have of the Financial Services systems. Those models are, of course, themselves generalisations that must be subjected to empirical testing using the 'Black Swan' logic. As discussed in Section III, adopting a full-blown functional perspective such as resilience engineering has major implications for the design of governance and control frameworks for Financial Services systems.

So what is the Financial Services system? Since the Financial Services system is a system like any other, it can be defined using the standard definition as a starting point: 'A system may be defined as a set of elements standing in interrelation among themselves and with the environment' (von Bertalanffy, 1975: 159). Ludwig von Bertalanffy, a biologist who devoted his lifetime to the development of the General Systems Theory, also introduced the idea that it is possible to describe 'the functioning of a system even if individual components are unknown' (von Bertalanffy, 1975: 121). This is obviously an important assumption given that Financial Services systems may be intractable. The primary goal of General Systems Theory was to develop principles for describing systems in which many variables, or forces, interact and to make the very organisation and dynamic of these forces the centre of attention of the study of system behaviour. Any attempt to reduce the behaviour of the overall system to the behaviour of individual elements, that is, to explain it by means of decomposition, will fail. General Systems Theory also introduced a distinction between two types of systems called closed and open, respectively. Closed systems are in equilibrium and are entropic. 'In going to equilibrium they typically lose structure and have a minimum of free energy; they are affected only by external "disturbances" and have no internal or endogenous sources of change; their component elements are relatively simple and linked directly via energy exchange (rather than information interchange); and since they are relatively closed they have no feedback or other systematic self-regulating or adaptive capabilities' (Buckley, 1968: 405). Open systems are negentropic, in the sense that they are can remain stable only because they can counteract a potential loss of control. They characteristically function to maintain their performance within pre-established limits. This requires feedback (and feedforward) loops to their environment, and possibly information as well as pure energy interchanges to serve the purposes of mainly self-regulation but also adaptation (change of system structure). The interchanges among their components may result in significant changes in the nature of the components themselves with important consequences for the system as a whole. Feedback control-loops enable self-regulation, as well as self-direction or at least adaptation to a changing environment, such that the system may change or elaborate its structure as a condition of survival or viability.

Most systems dealt with in physics can be categorised as closed systems with a defined and static structure. Examples of open systems include any biological system and, of course, any business system, that is, a system that must maintain a balance between assets and liabilities by a continuous exchange with its environment, that is, various types of markets. Clearly all Financial Services systems fall into the category of open systems and as a result have the following key characteristics:

- A continuous exchange between the system and its environments. A Financial Services system, or components of such a system, cannot survive without the constant exchanges on financial markets.
- A process of breaking down and rebuilding of components to maintain a desired system state. For example, a Financial Services system constantly needs to remove non-profitable elements and grow profitable elements.
- The presence of self-regulating processes. These processes are particularly important when the system pursues goals, regenerates components and reacts to severe 'disturbances'. In a Financial Services system, these self-regulating processes (that is, management processes) are driven by business goals and goals associated with regulatory requirements that are part of the environment in which the Financial Services system is operating.

While it might be impossible, and probably also of limited use, to describe each component of a Financial Services system in detail, it is quite possible to determine what types of elements the Financial Services systems should consist of. As any other socio-technical system (for example, Emery and Trist, 1965), Financial Services systems can be viewed as having organisational (social), human and technological components. Depending on the specific domain, that is, aviation, telecommunications, power generation and Financial Services industry, the coupling between these different types of components varies and has implications for the overall system behaviour and how we can get a better of understanding of why 'things go wrong'. As stated in the introductory comments, Financial Services systems are tightly coupled systems and actions taken in one part of the system may therefore affect other parts of the system in an unexpected and/or unpredictable manner. Only resilient Financial Services systems are able to withstand the impact of unexpected events. Their key characteristic, as resilient systems, is the ability to maintain functioning in both expected and unexpected conditions. The recent financial crisis illustrated that many Financial Services systems had become closed systems and lost their ability to self-regulate. Massive governmental actions in the form of bailouts were therefore required to temporarily stabilise the system.

Chapter 3
A Dynamic Systems Modelling Perspective

Gunilla Sundström and Erik Hollnagel

The previous chapter ended by concluding that Financial Services systems are open systems, and that they are dynamic, intractable and tightly coupled. We know from bitter experience that it is difficult to understand how Financial Services systems function. Because we cannot fully understand them they are difficult to control, and unexpected outcomes may therefore occur. The two main problems facing us are consequently how we can describe the behaviour of complex dynamic systems such as Financial Services systems, and whether is it possible to improve our understanding so that something similar to the events at the height of the 2007–2009 crisis do not happen again?

Everyday experience, supported by the philosophy of science, tells us that the possible explanations we produce depend on the perspectives we adopt. More formally, we do not see or perceive the world directly as it is, a philosophical position known as naïve realism, but only see what we are prepared to see and what we are capable of describing. This is well known in all sciences and has been described in a number of ways from the theory of linguistic realism (Whorf, 1940) to the *What-You-Look-For-Is-What-You-Find* principle (Hollnagel, 2009a). It also means that we are unable to manage or control that which we cannot describe. From a different perspective, cybernetics has formulated the Law of Requisite Variety, which simply states that the variety of the outcomes (of a system) can only be decreased by increasing the variety of the controller of that system (Ashby, 1956). The practical consequence of this law is that effective control is impossible if the regulator has less variety than the system, which happens if the controller does not fully understand how the system works (cf. Conant and Ashby, 1970: 95). Corresponding to that, Westrum (1993) has proposed a principle of requisite imagination or 'the fine art of anticipating what might go wrong' (Adamski and Westrum, 2003), as necessary to anticipate failures and unanticipated consequences. Requisite imagination clearly also requires an adequate understanding.

Chapter 2 introduced General Systems Theory and briefly discussed the difference between open and closed systems. This chapter will take a closer look at what cybernetics can offer to understand how open systems work. Cybernetics was literally created as a science in the late 1940s and formally presented in a book of the same name (Wiener, 1948). The subtitle of the book was 'the study of control and communication in the animal and the machine', and cybernetics has aptly been characterised as 'the art of ensuring the efficacy of action' (Wikipedia). In this chapter we will illustrate how we can leverage cybernetics to improve the understanding of how Financial Services systems function. In Chapter 10, we will describe how this understanding can be used to improve the resilience of Financial Services systems.

The 'Invisible Hand'

It has been known for a long time that the market, as an instance of what we now call a complex socio-technical system, sometimes seems to follow laws or forces that we do not immediately understand. In 1776, Adam Smith introduced the notion of the 'invisible hand' to describe the dynamics among self-interest (that is, the inclination of market participants to put their own interests first), competition (that is, the drive to increase market share), demand, and supply.

> But the annual revenue of every society is always precisely equal to the exchangeable value of the whole annual produce of its industry, or rather is precisely the same thing with that exchangeable value. As every individual, therefore, endeavours as much as he can, both to employ his capital in the support of domestic industry, and so to direct that industry that its produce maybe of the greatest value; every individual necessarily labours to render the annual revenue of the society as great as he can. He generally, indeed, neither intends to promote the public interest, nor knows how much he is promoting it.

> By preferring the support of domestic to that of foreign industry, he intends only his own security; and by directing that industry in such a manner as its produce may be of the greatest value, he intends only his own gain; and he is in this, as in many other cases, led by an invisible hand to promote an end which was no part of his intention. (Smith, 1999 [orig. 1776]: 32)

The hand is 'invisible' in the sense that the individual in the market does not understand what goes on, and therefore is unable to explain which actions give rise to which effects. But the hand was not invisible to Adam Smith, since he could describe how the market worked at the time. The hand is only 'invisible' if there is a lack of understanding of how the market works or if the understanding is limited or incorrect. If people falsely believe that they have completely understood something, for example the financial markets, they are naturally tempted to use this understanding to get the upper hand. This means that they inevitably venture into areas where their understanding is incomplete, and where there therefore is an 'invisible hand'. The folly is to think that there at any one time is a complete understanding, and also that any understanding – complete or incomplete – will remain valid forever. Markets – and the world – change, and an understanding that once was adequate will gradually become inadequate. There will therefore always be an invisible hand somewhere.

That the 'invisible' hand exists even today was amply illustrated when Alan Greenspan, who had been Chairman of the Federal Reserve of the United States from 1987 to 2006, in an interview by *The Guardian* (published 24 October 2008) admitted that he had 'discovered a flaw in the model that I perceived is the critical functioning structure that defines how the world works. I had been going for 40 years with considerable evidence that it was working exceptionally well.' As we all know by now, it was not.

If the hand is genuinely 'invisible', that is, if it is impossible to understand why things happen, then there is no chance of controlling the system. But cybernetics, in combination with control theory and General Systems Theory, provides the intellectual tools to make the 'invisible hand' visible, or in other words to develop an understanding of what goes on. The more specific set of tools is provided by a conceptual framework called the second cybernetics.

The Second Cybernetics

The basic way of regulating a process is through a feedback loop. This means that a measurement of the output of the system is fed back to the system and used as a basis for regulating the process that produces the output. The classical example is a thermostat, which regulates the temperature of a room by controlling heating (or cooling). If the measured temperature is too low, the thermostat switches on the heating unit; if the measured temperature is too high, the heating unit is switched off. The thermostat thus makes the system self-regulating by counteracting deviations in order to ensure that the room temperature stays within a desired range.

In economics, the simplest example is possibly the way in which central banks, such as the European Central Bank (ECB), triy to regulate the economy of a country or region. If the economy 'overheats', as measured by the inflation rate, the central bank tries to regulate the process by increasing the interest rate, as this is assumed to slow down the national economy. Conversely, if the economy 'cools', the interest rate is reduced, as this is assumed to stimulate the national economy. The system, meaning the national economy and the central bank, is self-regulating and deviation-counteracting – at least to the extent that the underlying model is correct.

This particular relationship is known as a negative feedback loop, meaning that the output from the regulator acts to oppose or attenuate changes to the input of the system. If the overall feedback of the system is negative, then the system will tend to be stable. The converse of that is a positive feedback loop in which

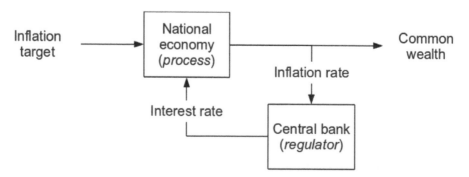

Figure 3.1   Control of the economy by negative feedback

the regulator responds so as to amplify the original signal instead of dampening it. An example of that is provided in the following.

While the example in Figure 3.1 is well suited to illustrate the principle of feedback control, it also makes it obvious that the model is much too simple. There clearly are several 'invisible hands' that affect how the national economy works – and perhaps even how central banks make decisions! Other ways of describing, or modelling, are necessary if we are to improve our understanding of the economy – and of complex, dynamic Financial Services systems in general.

A proposal for how to do this was described by Maruyama (1963), who called it 'deviation-amplifying mutual causal processes' or the 'second cybernetics'. (A similar approach was developed by Jay W. Forrester, who is best known for his work on system dynamics. One application to economics is Forrester, 1968.) Maruyama introduced the modelling principles as follows:

> By focusing on the deviation-counteracting aspect of the mutual causal relationships however, the cyberneticians paid less attention to the systems in which the mutual causal effects are deviation-amplifying. Such systems are ubiquitous: accumulation of capital in industry, evolution of living organisms, the rise of cultures of various types, interpersonal processes which produce mental illness, international conflicts, and the processes that are loosely termed as 'vicious circles' and 'compound interests'; in short, all processes of mutual causal relationships that amplify an insignificant or accidental initial kick, build up deviation and diverge from the initial condition. (Maruyama, 1963: 164)

The best way of explaining the principle of how 'deviation-amplifying mutual causal processes' work is to use an example from finance, taken from Sterman (2000). Figure 3.2 shows two coupled loops. The one on the left describes the couplings between relative value, demand, and price, while the one on the right describes the couplings between price, profits, and supply. The symbol ⊕ (and solid black lines) represents a proportional relation between two entities. For instance, if relative value increases, then demand also increases; and if relative value decreases, then demand also decreases. The symbol ⊗ (and dashed black lines) indicates an inverse relation between two entities. For instance, if price increases, then relative value decreases; but if price decreases, then relative value increases.

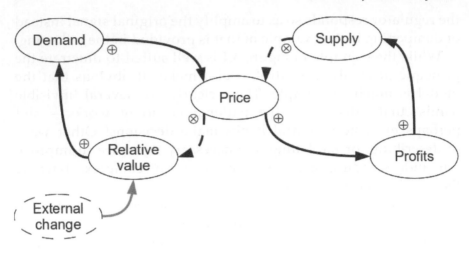

Figure 3.2   Mutually coupled processes

Source: After Sterman, 2000.

Making the 'Invisible Hand' Visible

The couplings represented by Figure 3.2 should be interpreted as follows. If some external change, such as a change in the price of substitutes or a change in confidence of valuation, increases the relative value, this will lead to an increase in demand, which in turn will lead to increases in price. Since an increase in price will lead to a decrease in relative value (due to the inverse relation between the two), the loop is stable. (The same reasoning, of course, goes if the effect of the initial external change was a decrease in relative value, or if the external change affected either of the two other functions.) An increase in price will at the same time lead to an increase in profits, which will lead to an increase in supply. Since this will lead to a decrease in price, which in turn will lead to a decrease in profits, and so on, the right-hand loop is also stable. Altogether there is an equilibrium between the two loops as well as in the system as a whole. An external change may destabilise it for a while, but after some time it will reach a new equilibrium. It is thus a clear example of an open system, according to the definition in Chapter 2.

By simply replacing the 'invisible hand' with two coupled multi-function loops, the understanding of how this Financial Services system works has been improved – to the extent that the five functions are sufficient to describe the system and that the coupling between them is correct. (There are also other assumptions, for instance that the changes propagate fast enough to dampen oscillations from external changes, and so on). We can then try to use the same principle to illustrate the possible dynamics of the 2007–2009 crisis in the Financial Services system.

One key assumption underlying the counterbalancing loops of the 'invisible hand' is that market participants have access to the same information, that the information is accurate, and that the market participants behave in a *rational* manner. This assumption was not met during the peak of the 2007–2009 financial crisis where a lack of information among lenders, borrowers and investors led to a loss of trust and eventually resulted in the complete freeze of financial markets. The observations that uncertainty about the quality of prospective purchases creates powerful dynamics in markets was suggested and formalised by Akerlof (1970). Figure 3.3 can be used to illustrate the impact of uncertainty, for instance a loss of confidence in valuations. If the confidence in valuations is reduced, this will reduce the relative value of an asset. This will lead to a decrease in demand and a decrease in price, which will have a negative impact on relative value because the confidence is low. A reduction in price will also serve to reduce the confidence even further. The left side of Figure 3.3 now illustrates a deviation-amplifying loop, which will end in an extreme state rather than an equilibrium. If the initial change is a reduction (loss in confidence of valuations), the end result will be a market freeze. And if the initial change is an increase (more confidence in valuations), the end result will be an unsustainable boom. In either case, the description of how deviations may be mutually amplifying allows us to understand how the 'invisible hand' works. The crucial difference between Figure 3.2 and Figure 3.3 is the change in the nature of the coupling between price and relative value from $\oplus$ (proportional) to $\otimes$ (inverse). A system with mutually coupled functions is generally stable if there is an odd number of inverse relations. The system on the left side of Figure 3.3 has an even number of inverse relations (in this case zero), and is therefore unstable.

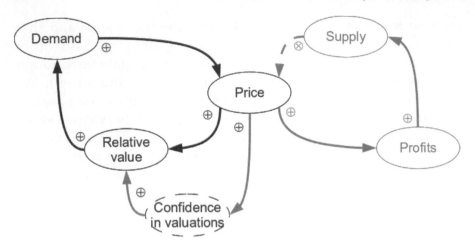

Figure 3.3   The impact of changes in the confidence of valuations

Pro-cyclicality

Another common perception of what happened during the 2007–2009 Financial Services systems crisis was that it was due to the combined impact of a credit boom and a housing bubble (for example, Richardson, 2009; Brunnermeier, 2009; Taylor, 2009; Shiller, 2009). Taylor (2009) found that credit booms and housing bubbles in the US seemed to be preceded by a longer period of low interest policy by the US Federal Reserve Bank. A low interest policy is, of course, usually associated with an increased supply of monetary resources. And, as illustrated in Figure 3.2, increased supply will lead to a decrease in price, that is, credit becomes cheaper. The Financial Stability Board (FSB), a key global body designed to 'address vulnerabilities and to develop and implement strong regulatory, supervisory and other policies in the interest of financial stability', published a report in April 2009 stating that pro-cyclicality was a major contributor to the emergence of the 2007–2009 financial crisis.

In the context of Financial Services systems, the term pro-cyclicality refers to the deviation-amplifying consequences of couplings between the global Financial Services system and the various economic sectors. In Figure 3.4, we provide an example of how such couplings might interact during the emergence of 'bubbles'. The lines represent couplings (feedback loops) that may

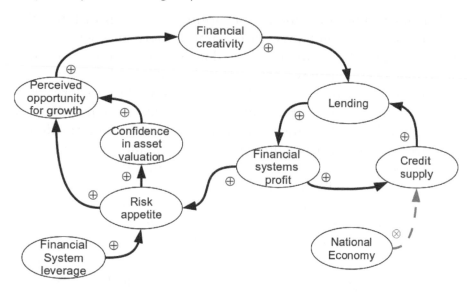

Figure 3.4   The destructive power of pro-cyclicality

be found in any Financial Services system, such as an individual Financial Services organisation. In addition to that there may be powerful couplings among the global and national economies, key stakeholders such as rating agencies, and Financial Services organisations (not included in Figure 3.4). Potential counteracting forces, such as capital inflows due to national monetary policies and/or Rating Agency ratings can be used to counteract mutual deviation-amplification, hence to dampen or stabilise the feedback loops of the Financial Services organisation. As Figure 3.4 shows, this Financial Services organisation (black solid lines only) is deviation-amplifying and the result will therefore be exuberance unless some counteracting force can dampen the deviations. In Figure 3.4, the national economy represents such a counteracting force, since it can be used to 'strangle' the credit supply. Shiller (2008:41) argues that it is critical to understand the '*social contagion* of boom thinking', facilitated by the factual observation that prices are rising. In fact, Shiller's social contagion is very similar to the previously mentioned confirmation bias (cf. Klayman and Ha, 1987), the powerful psychological process that makes us look for and interpret information in alignment with an existing belief. In the case of bubbles, the belief is simply that asset valuations will continue to appreciate.

The idea to use cybernetics to capture behaviour of Financial Services systems is not novel (for example, Sterman, 2000; Åström and Murray, 2008). However, to date there are few examples of frameworks that have taken full advantage of the dynamic system view and of the second cybernetics to create approaches for governance and control of Financial Services systems. We will return to this point in the book's third section. From a resilience engineering perspective, this chapter has made it clear that a resilient system must have the ability to counteract the impact of 'deviation-amplifying mutual causal processes'. To do this a system must be able to analyse and represent its state and the impact of actions on its future states. The ability to proactively anticipate future states is a critical capability for a resilient system.

While the second cybernetics indicates how a basis for resilience can be developed, the approach also has some shortcomings. The most serious is that it describes the functions and the couplings at the same time, so that the 'architecture' of the systems is given a priori. Another shortcoming is that the couplings are represented by a single relation, which can be either proportional or inverse. But any function may be coupled to other functions in more complex ways. Resilience engineering can, however, provide a solution to these shortcomings, as will be described in Section 3.

Chapter 4
From the Efficient Market Hypothesis to Econophysics

Bill McKelvey and Rossitsa Yalamova

With the failure of LTCM (Long-Term Capital Management) in 1998, Nobel Laureates Myron Scholes and Robert Merton learned the hard way about the limitations of the 'value-at-risk' (VAR) models. As described in *The Economist* (2009a: 71), VAR models are:

> used by institutional investors to work out how much capital they need to set aside as insurance against losses on risky assets. These models mistakenly assume that the volatility of asset prices and the correlations between prices are constant, says Mr Scholes. When, say, two types of asset were assumed to be uncorrelated, investors felt able to hold the same capital as a cushion against losses on both, because they would not lose on both at the same time.

The observation made in the *The Economist* was that the LTCM debacle and the 2007–2009 financial crisis both demonstrated just how quickly supposedly 'uncorrelated' asset-price movements could become highly correlated and lead all traders toward the same buy/sell motivations ending in a market crash. This phenomenon was, however, not new. In the beginning of the twentieth century, Bachelier invented the random walk in an attempt to describe price fluctuations on the Paris stock exchange as part of his dissertation (Bachelier, 1914). His advisor, Henri Poincaré, in reviewing Bachelier's dissertation urged caution in applying the principle to human behaviour, and wrote:

> When men are brought together they no longer decide by chance and independently of each other, but react upon one another. Many causes come into action, they trouble the men and draw them this way and that, but there is one thing they cannot destroy, the habits they have of Panurge's sheep. (Poincaré, 1914: 88)

(The reference to Panurge's sheep is taken from Rabelais' book *Gargantua und Pantagruel*, and alludes to the phenomenon that stock traders seem to behave in the same manner as the sheep that blindly jumped overboard after Panurge had thrown the alpha ram off the ship. Clearly not a very desirable behaviour from any perspective.)

Since this kind of correlated, or herding, behaviour seems to lead to crashes, the key question is how financial market behaviours can be made resistant against this. Some, such as Myron Scholes, suggest establishing independent capital reserves for different asset types to avoid unexpected crash-creating correlated asset-price movements leading to crashes (*The Economist*, 2009a). From our perspective such moves are not necessary if markets are in a 'healthy' state, that is, in efficient-market mode (Fama, 1970) where greed, fear and uncertainty are balanced. On the other hand such moves will be insufficient if markets are in an 'unhealthy' state where market behaviour consists of primarily correlated trading of highly leveraged financial products.

Our perspective is based on key elements of *Econophysics* (West and Deering, 1995; Mantegna and Stanley, 2000; Vasconcelos, 2004), Minsky (1982, 1986), power-law science (Andriani and McKelvey, 2010), and log-periodic power laws (Yan, Woodard and Sornette, 2010). Leveraging these concepts, we propose that it is possible to proactively identify (that is, anticipate) the tipping points where efficient markets transition into crash-producing stock-market bubbles. A resilient system must have the ability to anticipate such market tipping points. Without this, a Financial Services system will suffer from the impact of a bubble without being able to pre-empt the likely very negative impact this may have on its own performance.

The Economist (2009c: 70) made the observation that the 25-year '*Great Moderation*' period has ended. In this period, macro and financial economists typically presumed that their theoretical and quantitative models offered indisputably correct views of real-world market behaviour. Following the example of Bachelier, these models embraced independent and identically distributed (*i.i.d.*), normal distribution assumptions of efficient market traders' random walks and asset-pricing behaviours. Yet the 2007–2009 liquidity crisis, banking failures, and consequent

'Great Recession' of uncertain duration remind us that it is still possible for unanticipated *correlated behaviours* suddenly to turn into extreme events. We propose to leverage the principles of econophysics to make Financial Services systems more resilient and thereby minimise the damage from rare but inevitable extreme volatilities and market crashes. In particular, we argue that it is critical to intervene at, or even slightly before, the point when traders begin to show self-organised herding behaviour. At this point a bubble build-up is very likely and interference with this build-up will lead to a faster and less painful return to the efficient market behaviour presumed by Fama's efficient market hypothesis.

The Efficient Market Hypothesis (EMH; Fama, 1970) is one of the key paradigms in finance together with the Capital-Asset Pricing Model (CAPM; Sharpe, 1964; Lintner, 1965; Black, 1972), and the Options-Pricing Model (OPM; Black and Scholes, 1973). The EMH states that the actions of the many competing participants in the market will cause the actual price (of a security) to wander randomly about its intrinsic value (Fama, 1965). The EMH is, however, very controversial and hotly disputed. For example, Cooper (2008: 11) notes that 'Despite overwhelming evidence to the contrary, the Efficient Market Hypothesis remains the bedrock of how conventional wisdom views the Financial Services system'. The recent crisis is another example of how the EMH view is contradicted by real world events, such as the emergence of 'irrational exuberance' (Greenspan, 1996) on financial markets. Mandelbrot and Hudson (2004: ix) state that '"Modern" financial theory is founded on a few, shaky myths that lead us to underestimate the real risk of financial markets' and 'Orthodox financial theory is riddled with false assumptions and wrong results' (Ibid.: x). However, Fama (1998) maintains that until new and better paradigm(s) are put forth, one cannot criticise EMH/CAPM. Fama reduces behavioural finance – and trading dynamics – to anomalies and over-/under-reaction episodes that are normally distributed.

Key Elements of Econophysics

Mirowski (1989) achieved notoriety in economics for his book, *More Heat than Light*, in which he described economics as a kind of thermodynamic-equilibrium-based 'social physics'. Economists have always been attracted to physics from the time of Adam Smith's *Wealth of Nations* (1776). Boltsmann's creation of statistical mechanics in 1877 was used in late nineteenth-century marginalism, and Marshalian neoclassical economics referred to the concept of a thermodynamic equilibrium. Bachelier's invention of the random walk gave rise to faith in the 'average' stock price and Gaussian statistics (Fox, 2009). No one has, however, tried harder to make economics a physics-like science than Samuelson. His goal was to reduce the logic of neoclassical economics to mathematical equations. This culminated in his book of 1947, *Foundations of Economic Analysis*, which built from the physics equations of thermodynamics and random walks.

There is, however, one crucial difference between economics and physics, namely that theories in physics are proven by empirical reality (although it sometimes takes a long time for truth to emerge). In economics there is much less evidence that the economists' faith-based micro- and macro-economics mathematics actually corresponds to reality. A few examples illustrate this point. For a more concise discussion the reader is referred to Rosser (2008).

- 'No economic theory was ever abandoned because it was rejected by some empirical econometric test' (Spanos, 1986: 660).
- 'It seems to me that as a profession we don't value empirical work very highly' (Leamer, 1990: 180). 'I believe there are very few empirical enterprises in economics that have any significant possibility of affecting how economists think about how economies operate' (Ibid.: 193).
- 'We don't genuinely take empirical work seriously in economics. It's not the source by which economists accumulate their opinions, by and large' (Leamer, quoted in Kennedy, 2003:7).

- 'I invite the reader to try and identify ... a meaningful hypothesis about economic behaviour that has fallen into disrepute because of a formal statistical test' (Summers, 1991: 139).

The field of econophysics differs from traditional economics first by having a strict focus on empirical findings based on large databases and second by a shift in emphasis from marginalism and Gaussian statistics to Lévy and Pareto distributions and scalability.

The term econophysics was introduced by Stanley in 1995 at a conference on statistical physics, although early discoveries date as far back as Pareto (1897), Auerbach (1913), and Zipf (1929). Under the label econophysics, researchers have listed a large number of findings about long-tailed Pareto-distributed economic phenomena. Instead of assuming that economic phenomena invariably trend toward equilibrium, means, and finite variance, we see all kinds of findings to the contrary, including (cf. Rosser, 2008):

- Returns in financial markets (Mantcgna, 1991; Wong et al., 2009).
- Economic shocks and growth rates (Canning et al., 1998; Redelico et al., 2008).
- Firm size and growth rates (Stanley et al., 1996; Zhang et al., 2009).
- Income and wealth (Levy and Solomon, 1997; Yakovenko and Rosser, 2009).
- City sizes (Rosser, 1994; McKelvey, 2011).
- Networks (Barabási, 2002; Song et al., 2009).
- Stock market volatility dynamics (Ghashghaie et al., 1996; Jiang et al., 2010).
- Detecting stock market bubbles and predicting crashes (Hurst, 1951; Varkoulas and Baum, 1996; di Matteo et al., 2005; Alvarez-Ramirez et al., 2008; Czarnecki et al., 2008).

The differences between classical economics and econophysics have aptly been explained in the following paragraphs from a recent article by Yan et al. (2010):

The research presented here is highly unconventional in financial economics and will swim against the convention that bubbles cannot be diagnosed in advance and crashes are somehow inherently impossible. But as Einstein once said: 'Problems cannot be solved at the same level of awareness that created them.' We thus propose a kind of Pascal's wager: Is it really a big risk for the community to explore the possibility of changing the conventional wisdom and open new directions for the diagnostic of bubbles, *ones that may eventually lead to important policy and regulatory implications*?

This research may have indeed global impacts. We confront directly the widespread belief that crises are inherently unpredictable. If one can convince that some crises can be diagnosed in advance and, what is even more important, if one can quantify the associated uncertainties, *this may help economists and policy makers develop new approaches to deal with financial and economic crises*. (Italics added)

Later in this book both Chapters 6 and 7 focus on how to make systems more resilient by developing the capability to anticipate. Each chapter builds directly on the existing econophysics work focusing on volatility dynamics, multifractality, and the so-called Hurst exponent.

Econophysics and Resilience Engineering

Econophysics can be seen as an instance of complexity science in the way it studies the scalable outcomes of emergent phenomena (Brock, 2000; Gell-Mann, 2002). While the econophysics findings are mostly empirical, the theoretical basis for power-law distributions is based on the work of Mandelbrot (1963a, 1963b; 1982), as well as the more recent works of Schroeder (1991), West and Deering (1995), Mantegna and Stanley (2000), Vasconcelos (2004), and Newman (2005). This literature describes how order emerges once the forces of self-organising agents – such as biomolecules, organisms, people, or social systems – are set in motion. In his opening remarks at the founding of the Santa Fe Institute, Gell-Mann (1988) emphasised the search for scale-free theories – simple ideas that explain complex, multi-level phenomena. Brock (2000) went so far as to say that *scalability* was the core of the Santa Fe vision – meaning that no matter what the scale of measurement is, the phenomena appear the same and result from the same causal dynamics (cf. Gell-Mann, 2002).

More precisely, econophysicists apply concepts of statistical physics to Financial Services systems. When it is not possible to write an equation to explain the dynamics of all the relatively 'microscopic' entities of a Financial Services system, econophysicists apply scaling concepts to explain macro-level dynamics without having to first create the equations of 'microscopic' interacting entities. Scaling concepts may thus be suitable for the modelling of intractable systems that cannot be described in detail (see Chapter 2). Econophysics seems especially useful for describing how stock-trading dynamics shift from the EMH assumptions to become embedded in the development of a bubble build-up to the crash point (Jiang et al., 2010; Zhou and Sornette, 2003; Yan, Woodard and Sornette, 2010).

The key theoretical concepts on which econophysics is based are *fractal structures, power laws,* and *scale-free theory.*

- *Fractal structures.* The defining property of a fractal structure is self-similarity. Consider, for instance, the cauliflower. Cut off a 'floret'; then cut a smaller floret from the first floret; then an even smaller one; and then another, and so on. Despite getting smaller and smaller, each floret has roughly the same design as the one 'above' and 'below' and performs the same function. Fractals can result from mathematical formulas – as shown in Mandelbrot (1982). Econophysics is more interested in fractal structures that come from adaptive processes in biological, social, and financial contexts, where the same adaptation dynamics appear at multiple levels. McKelvey et al. (2011) cite 19 studies showing adaptation-based predator/prey fractal dynamics, while Zanini (2008) argue that the same effects hold for merger and acquisition activities in business niches. Fractal structures are often indicated by power laws. Barabási (2002) connects scalability, fractal structure, and power-law findings to social networks and show how networks in the physical, biological and social worlds are fractally structured and that there is a *'rank/frequency'* effect – an underlying Pareto distribution showing many sparsely connected nodes at one end and one very well connected node at the other.
- *Power laws.* A Pareto rank/frequency distribution plotted on double-log scales appears as a power-law distribution – an inverse sloping straight line. Power laws often take the

form of rank/size expressions such as $F \sim N^{-\beta}$, where F is frequency, N is rank (the variable) and β, the exponent, is constant. However, in a typical 'exponential' function, for example, $p(y) \sim e^{(ax)}$, the exponent is the variable and e is constant. The now famous power law 'signature' dates back to Auerbach (1913) and Zipf (1929, 1932, 1949). Andriani and McKelvey (2007, 2009) listed about 140 kinds of power laws in physical, biological, social, and organisational phenomena. Both Stanley et al. (1996) and Axtell (2001) found that manufacturing firms in the US showed a fractal structure. Since power laws mostly appear to be the result of self-organisation, they often, if not always, signify active processes that maintain some kind of Bak's (1996) self-organised criticality. Thus, Ishikawa (2006) shows power laws in adaptive and changing industries (as opposed to static ones), while Podobnik et al. (2006) show power laws in the stock markets of transition economies. The Dow Jones market capitalisations of the 30 largest US publicly traded firms show a power law – again, evidence of fractals when traders were free to buy and sell as they wish (Glaser, 2009).

- *Scale-Free Theories* explain why fractals appear and behave as they do. Though scalability may have been at the core of the Santa Fe vision (Bak et al., 1987; Bak, 1996), scale-free theories have only recently been consolidated and used collectively by the econophysicists (West and Deering, 1995; Mantegna and Stanley, 2000; Vasconcelos, 2004; Newman, 2005). The key feature that sets scale-free theories apart from most social science theories is that they use a single cause to explain fractal dynamics at multiple levels. An early example is Galileo's *Square-Cube Law* (formulated in 1638); the cauliflower illustrates this as it keeps subdividing to keep its surface area at a constant ratio to its growing volume. Explanations for why some structures have adaptive success while others do not, range from biology to social science. Andriani and McKelvey (2009) describe 15 scale-free theories applying to firms.

In Section II of this book, Chapter 5 uses concepts from econophysics to describe the short history of financial engineering starting with the invention of derivatives, laptop computers and mortgage-backed securities. In the chapter, we suggest that financial-engineering really facilitated the emergence of the 2007 liquidity crisis. We also illustrate how the use of scale free theories might have helped to highlight the preconditions that eventually scaled up into the Dow Jones crash from a high of 14,198 down to 6,469.

In Section II of this book, we also describe some of the theory and research that allows us to point to the *tipping point* at which stock markets switch from EMH random-walk trading to herding behaviours (Hirshleifer and Teoh, 2003) to become power-law distributed bubble build-ups based on what is termed the 'volatility autocorrelation function', and which is signified by the Hurst exponent. The tipping point is when we believe that resilient systems need to be prepared to take action. We also suggest a set of resilience devices ranging from some aimed to thwart 'reckless endangerment' by Financial Services firms to others set in motion by various elements of 'exuberant' increases in risk-taking.

SECTION II
Understanding What: Making Sense of Unpredictable Events and Developments

Introduction

Erik Hollnagel and Gunilla Sundström

In the Introduction to Section I, we stated that we must go beyond a simple sequence of events ordered on a timeline to make sense of how the recent financial crisis emerged. In subsequent chapters, we described financial systems as complex socio-technical systems. And most importantly, we recognised that the focus must be on creating the capabilities of the system to 'make sense' of unexpected events and adjust what they do so that system performance can be sustained over time, in line with the concept of resilience.

The study of human perception has taught us that what we see is not simply a reflection of the physical stimuli we are exposed to. For example, the perceived size of the moon seems larger when it is at the horizon than when it is high in the sky, even though we know that the actual size of the moon is the same. What has changed instead is our frame of reference. Other illustrations of how perception simply is not a reflection of the arrangement of physical elements are visual illusions. For example, in Zöllner's illusion (Weisstein, 2010) four vertical lines seem to be tilted due to the positioning of crossing lines, even though the four lines are parallel. Moreover, even though we know that the lines are parallel, we still perceive them as tilted. As a consequence, 'making sense' requires us actively to go beyond immediate

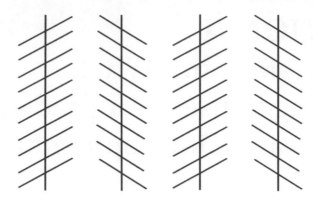

Figure II.1   Zöllner's illusion

perceptions and to look for answers. And then, we are of course back to the WYLFIWYF principle introduced in Chapter 1.

The study of organisational sense-making has identified the importance of focusing on the 'interplay of action and interpretation' (Weick and Sutcliffe, 2005: 409), rather than on the evaluation of options. The latter is of course the primary focus of formal decision-making theories and much related to the assumption of the *homo economicus*. We believe that understanding the interplay of action and interpretation is particularly important in complex financial systems for the simple reason that their behaviour relies on interpretation of information from a complex set of sources. Those perceived as more credible than others are more likely to have more influence on sense making. For example, if a national government reports stellar growth, we are likely to add this information to our already strong belief that real estate assets will continue to appreciate. Our investment actions are likely to be aligned with this expectation. The key question is then how do expectations emerge? What role do events play in resetting expectations? Is it possible to detect such changed expectations? How does financial systems' behaviour change as a result of changed expectations? All three chapters in this section are focused on providing answers to one or more of these questions.

In Chapter 5, McKelvey and Yalamova suggest that we can make sense of the events leading to the 2007–2009 financial crisis leveraging three concepts: 'tiny initiating events' (TIEs), scale-free theories and positive feedback loops. TIEs by themselves might often be perceived as insignificant and independent events; however, when placed in the context of scale-free theories with a focus on system interdependencies, it becomes clear that TIEs easily can trigger spirals leading to potentially catastrophic outcomes. The ability to make sense out of the TIEs is therefore necessary for a system to adjust its actions in order to sustain performance.

In Chapter 6, Satinover and Sornette point out that designers of financial instruments usually fail to recognise that financial markets are tightly coupled complex systems. As a result, they expect that events are independent, when in fact they are connected, and as a result they underestimate the risk. Satinover and Sornette illustrate this with some simple examples in their chapter. A key message in the chapter is that it is important to be able to identify the so-called Dragon Kings. These are events that are driven by a different system dynamic and therefore are important precursors of a phase transition in a system, for example, the transition from a healthy financial market state to an unhealthy financial market state. The latter is characterised by increased connectivity among previously independently acting agents, that is, traders in a stock market. Finally, a methodology is proposed for proactively predicting financial bubbles; the methodology is briefly highlighted in Chapter 6.

Finally, in Chapter 7, Yalamova and McKelvey propose a phase transition model to characterise healthy and unhealthy financial market behaviour. A healthy market is assumed to behave in alignment with the Efficient Market Hypothesis, whereas an unhealthy market is described as market in which trading behaviour is highly connected and best described as 'herding' behaviour. Most importantly, the authors suggest that a metric called the Hurst exponent can be used to detect when the financial market starts transitioning into an unhealthy state, thus creating a need to change expectations about future market behaviour. It is of course at this point that a resilient system needs to start taking actions to maintain a healthy state. How to improve system resilience is the primary focus of Section III.

Chapter 5

The 2007 Liquidity Crisis: An Example of Scalability Dynamics in Action

Bill McKelvey and Rossitsa Yalamova

In Chapter 4 we gave a brief introduction to econophysics in general and illustrated how some key concepts have been applied to stock markets. This discussion also pointed out that Mandelbrot's (1963a; 1997) work on fractals was the precursor to the power-law aspects of econophysics. In the present chapter, we take a closer look at some financial instruments dating back to 1972, specifically *derivatives*, computer-based *financial engineering*, and *mortgage-backed securities*. These financial instruments were important precursors to the 2007 liquidity crisis and the following worldwide recession. We will refer to this as the Great Recession in analogy with the Great Depression of the 1930s. This chapter sets the stage for our identification, in Chapter 7, of the tipping point where Fama's (1970) efficient markets are transformed into crash-producing stock-market bubbles.

There have been both small and worldwide crashes since 1972 such as the 'Black Monday' of 19 October 1987; the Asian financial crisis of 1997; the Russian financial crisis of 1998 (which led to the failure of Long-term Capital Management and the US Federal Reserve's first rescue plan); and the 'dot.com bust' of 2002. Although the 2007 liquidity crisis was severe enough to set off the Great Recession, the Great Recession was not caused by any single internal or external event to the US Finance industry. Instead, it was the result of the changes in Financial Services products, services and regulations in the US from 1972 to 2007;

the impact of these changes eventually propagated through the global Financial Services system. Examples of these changes include derivatives, financial engineering, mortgage-backed securities, the repeal of the US Glass-Steagall Act, the interplay between fixed-interest subprime mortgages and variable rate mortgages, and so on. All these changes and innovations eventually combined in 2007 to trigger the Great Recession. Because of the connectivities between the buy–sell behaviours of underlying banks and traders, we see the emergence over time of multiple skewed distributions in stock-market price volatilities. We contend that no single scale-free phenomenon is sufficient by itself to set of a major market crash.

We note, however, that one of the scale-free theories, 'combination theory' (Newman, 2005; cf. Table 5.1), proposes that when several skew-distributed variables interact, the most likely consequence is a power-law (or Pareto) distribution with an extreme outcome. This means that although no single spiralling process or underlying scale-free dynamic set off the liquidity crisis, their combination from the invention of derivatives in 1972 to the crash of 2007 built up something that ended in the crisis and the post-2007 Great Recession.

Scale-free theories (SFTs) have already been introduced in Chapter 4 to explain the various kinds of connectivities, or couplings, among banks which set the scalability dynamics in motion. Another important concept is that of *tiny initiating events* (TIEs; Holland, 1995, 2002). These are defined as tiny and seemingly insignificant events which nevertheless may have exceedingly large consequences, due to the non-linear nature of the system. The classical example of that is the butterfly effect (Lorenz, 1972). The TIEs are seen as triggering 'spiralling causal dynamics' as they are described by the SFTs.

Identifying Scalability Dynamics from 1972 to 2007

In this section we describe the shift from Bachelier's (1914) random walk and Fama's (1970) hypothesised efficient market behaviours to stock-market price volatilities based on positive feedback processes. In Table 5.1, we point to six specific SFTs drawn from Andriani and McKelvey (2009) that offer different types of explanations about spiral developments; we associate these SFTs with the various dates we identify in the progression toward 2007.

Two articles in *The Economist* (2009d, 2010) acknowledge that bank trading does not follow Bachelier's (1914) random walk and Fama's (1970) efficient market hypothesis. Economists traditionally make an assumption about independent actions among rational actors (by statisticians called *i.i.d.* – independent and identically distributed behaviour), but *The Economist* articles portray a series of bank-to-bank connectivities, that is, couplings, leading to the liquidity crisis of 2007:

> Banks mimic other banks. They expose themselves to similar risks by making the same sorts of loans. Each bank's appetite for lending rises and falls in sync. What is safe for one institution becomes dangerous if they all do the same, which is often how financial trouble starts. The scope for nasty spillovers is increased by direct linkages. Banks lend to each other as well as to customers, so one firm's failure can quickly cause others to fall over, too. (*The Economist*, 2009b: 88)

The article points out what most economists and Fama's efficient market disciples do not see: trading connectivities among banks in the US and worldwide that result from the scalabilities discussed in Chapter 4. Like anyone else, economists cannot see what they not are looking for. The transition from efficient market behaviours to scalable behaviours will be discussed in greater detail in Chapter 9. At this point we will describe the idea of positive feedback processes, Minsky (1982, 1986), as well as a broader set of SFTs. This should give the reader a better understanding of what we have in mind when looking for the TIEs, Spirals, and SFTs in our Liquidity Crisis Example. We focus particularly on the two key elements that usually have to be present for scalability to develop: tension and connectivity.

Positive Feedback and Minsky Moments

Although 'deviation amplification' and positive feedback dynamics date back to the classic insights of Maruyama (1963; cf. Chapter 3), Minsky brought the idea of positive feedback into stock trading as an alternative to Bachelier's (1914) random walk and Fama's (1970) hypothesised efficient-market behaviour. One marker of this is the now famous phrase: *stability creates instability*

(Minsky, 1982, 1986). Lahart (2007) uses the 'Minsky Moment' to describe the inflection points where positive-feedback forces stock markets up in value and then reverses to force them down. Wolfson (2002) summarises Minsky's perspective in terms of the four key positive feedback cycles (cf. Table 5.1).

Table 5.1   The four feedback loops

	Positive Feedback (PF) up	'Minsky Moment' PF down
Financial fragility	Optimism increases; attitudes toward debt and risk change; liability structures change; the Financial Services system becomes increasingly fragile; as debt increases, fragility increases.	As short-term debt increases, liquidity declines; speculative (Ponzi) firms increase; liquidity decreases and interest rates rise; firms increase short-term debt in order to pay off long-term debt.
Movement to the brink	As the speculative bubble increases, the US Federal Reserve, and other central banks increase interest rates to slow it down. Speculative hedge funds, banks and Ponzi firms have to keep increasing debt as interest rates increase.	The US Federal Reserve raises its discount rate from a low of 0.75 per cent in November 2002, and sets off the financial crisis by overly increasing the debt load of all those speculating: higher interest → more debt expense → more borrowing → higher interest, and so on.
Surprise event	As Financial Services systems become more speculative they become more vulnerable to 'unusual' events (the failure of a large bank, bond default by a foreign country). The more financial fragility the more likely surprise failures.	The more failures the more financial fragility → the more fragility the more failures → etc.
Debt-deflation	Financial fragility increases unwillingness to finance investment; less investment reduces profits, which reduces willingness to invest. Decline in profits leads to debt-deflation, which reduces prices, which in turn increases the real value of outstanding debt commitments.	The US Federal Reserve (and US government) tries to stimulate the economy. This in turn sets off an inflationary spiral which sets the stage for another round of inflationary expansion: reduced interest rates → increased borrowing → increased house prices → mortgages go down toward subprime levels → housing speculation starts → house-based bubble starts → etc.

The table opposite shows that there are several self-reinforcing positive-feedback processes that can cause bubble build-ups in stock markets that can lead to dramatic crashes. They begin with initiating events such as *new trading rules, hedging techniques,* or the *development of new derivatives-based products*. Individual traders and institutions engage in these initiating events and in the process of learning and copying, herding behaviour begins (Brunnermeier, 2001; Hirshleifer and Teoh, 2003) and a bubble-build-up develops.

Tension and Connectivity

Bak (1996) recounts his 1987 study of change dynamics in sandpiles. Falling sand grains are allowed to slowly pile up. Eventually the sandpile becomes so high and the sides of the sandpile so steep that sand grains start sliding down its sides to keep readjusting its height and the steepness of its sloping sides to maintain stability. Bak called this *self-organised criticality* – the sliding sand grains keep the slope at the same *critical* angle. The degree of steepness of the slope depends on two elements: (1) *gravity* and (2) the *sharp irregularity* of the grains. Take away gravity and the grains will not slide down the side of the sandpile. Take away irregularity, for example, shift from falling irregular grains of sand to, say, *M&M Peanuts,* and the M&Ms will not pile up (try it and see what happens!). Bak also discovered that the size of sand grain movements ranged from frequent instances of one or a few grains moving to small avalanches to a rare rather large avalanche, and that the size and frequency of falling grains was power-law distributed!

The analogy with the sandpile can be used to understand the financial build-up phenomena before the financial crash:

- *Gravity* can be replaced by *greed, liquidity, risk taking,* and *arrogance,* which all force people to invest to make profits.
- *Sharp irregularity* can be replaced by human connectivity, for instance by *phones, email,* financial engineers gathered in *one office area* in a firm, quant types *transferring* from one firm to another, advertising and *advice-giving* by firms, and especially *human copying, herding, and learning abilities,* and general interest in learning better approaches for making money.

This analogy reveals that that the two critical elements necessary for power-law distributions – tension and connectivity – are present in stock market trader behaviours. The many bases of tension coupled with the many bases of connectivity mean that financial market trading can be characterised by the four key positive-feedback cycles described in Table 5.1, as well as several other SFTs.

Scale-Free Theories Relevant for the 2007 Crisis Build-Up

A key feature that sets SFTs apart from many social science theories is that they use a single cause to explain fractal dynamics at multiple levels. More specifically, when tension and connectivity are present among a set of agents (stock traders in the following example), it is likely that one or more TIEs will spiral up leading to an extreme outcome such as a market crash, that is, a bubble-build-up occurs. We can point to many 'spiral' examples dating back to 1972.

While *tension exists all the time*, resulting from the need to make money or from the fear of losing money, the *connectivities* giving rise to herd behaviours come and go, and can change their nature. Explanations why some structures have adapted successfully via self-organised criticality while others do not, range from physics to biology to social science.

Andriani and McKelvey (2009) describe 15 SFTs for organisations and business activities. In the following, we describe six of these briefly and how they played a role in the build-up to the 2007 liquidity crisis (see Table 5.2). This makes it clear that there was not just one SFT and not just one point in time that set off the crisis. Many scalability spirals were set in motion after the invention of derivatives in 1972, and the combination of them was sufficient to alter the institutional structure of Finance in the US, which resulted in the crisis and the Great Recession worldwide.

Table 5.2 SFTs associated with the crisis build-up

SFT #1: Phase transition (Turbulent flows)

External energy above a specific threshold (called the 1st critical value) may cause phase transitions in which internal energy flows exhibit self-reinforcing positive feedback. In combination with interaction effects from tiny initiating events, this may spread throughout a system so that new interaction groupings form with a Pareto distribution of size from smallest to largest (Prigogine, 1955; Nicolis and Prigogine, 1989).

1972	1st critical value threshold lowered; use of derivatives lowers the 1st critical value threshold such that high-risk a high-leverage investment practices emerge with less initial risk.
1979	1st critical value threshold lowered (computers become cheaper; 'quants' move into Finance); emergent degrees of freedom in re-combinations of high-risk tranches are obscured by the complexity of the securitisation packages.
1986	Securitisation packages lowered the risk threshold for taking advantage of Mortgage-backed-securities (MBSs); what emerged was a complex obscuritisation of mortgage-backed risk vulnerabilities.
2000	1st critical value threshold lowered; fosters emergent low-interested based mortgage policies.

SFT #2: Contagion bursts (Epidemics; idea contagion)

Viruses, stories, and metaphors often spread exponentially – each person 'infects' two others and the effects expands geometrically. Changing settings, such as almost empty or very crowded rooms and airplanes, will change rates of contagious flow leading to bursts of contagion or spreading via increased interactions. These avalanches result in the power-law signature (Watts, 2003; Baskin, 2005) because of the 'small-world' structures of the underlying networks.

1972	Banks substitute for crowded rooms, buses, and planes, thereby spreading contagion more quickly.
1999	Contagion is bank-by-bank as opposed to via random individuals; speeds up gross risk-taking by banks.
2000	Banks and lenders stop asking for credit and income statements; borrowers learn about cheap mortgages; towns and builders learn to build housing developments to take advantage; real estate agents start speculative buying, reselling.
2002	Agents connected to banks speeds up the spread of the teaser loan idea.
2006	All the connectivities among banks now foster rapid decline rather than rapid growth.

SFT #3: Irregularity generated gradients (Coral growth; blockages)

A random, insignificant irregularity (e.g., a tree falls into a stream to create a blockage), coupled with positive feedback (more leaves, plants, mud, debris, etc., collect at the tree making the blockage larger), may start a self-reinforcing (autocatalytic) process that creates a niche in which plants and animals can grow. This process explains the growth of coral reefs, innovation systems and the emergence of creative new products (Turner, 2000; Odling-Smee et al., 2003).

Table 7.2 continued   SFTs associated with the crisis build-up

1979	Individually insignificant innovations plus positive feedback fuel rapid, unwatched, and unchecked growth.

SFT #4: Spontaneous order creation (Heterogeneous agents)

Seeking other agents to copy/learn from to improve fitness generates networks; some networks may in turn become groups, some groups may form larger groups and hierarchies (Kauffman, 1993; Holland, 1995).

1979	Banks, acting as agents, communicate, learn, influence each other with positive feedback effects.
2000	People learn to use cheap money to refinance their houses; often several times; mortgages increase in value; debt-based spending increases; fragility of the economy increases.

SFT #5: Combination theory (Number of exponentials; complexity)

Multiple exponential or lognormal distributions, or increased complexity of components subtasks, processes) may lead to outcomes that follow a power-law distribution (West and Deering, 1995; Newman 2005).

1986	Increased complexity of securitisation packaging results in combinations of high-default-risk tranches, that compound into securitisation packages more likely to show skew distributions and Pareto extremes.
2003	Increased complexity of securitisation packaging with increasing leverage results in combinations of skew distributions that compound into Pareto extremes of vanishing bank liquidity (that is, loss of bank liquidity in amounts never experienced previously in the US nor worldwide).

SFT #6: Preferential attachment (Nodes; gravitational attraction)

When new agents arrive to a system, larger nodes have an enhanced propensity to attract agents and will therefore become disproportionately even larger, resulting in the power law signature (Barabási, 2002; Newman, 2005).

2002	Some US lenders and banks (example.g., Countrywide, IndyMac, Washington Mutual, Freddy Mac, Fanny Mae, etc.), became especially well known for offering teaser loans with few questions asked; they then had links to other banks (which were often later acquired).
2006	'Attraction' in reverse; i.e., the networks and attachments now spread panic and decline.

Analysis of the 2007 Liquidity Crisis

In the following account of developments leading up to the August 2007 liquidity crisis, we try to identify relevant TIEs, the follow-on growth spirals leading up to the *Minsky Moment's,* and the SFTs that best identify and explain the various ways traders' tensions and connectivities caused the eventual market crash.

The technical details are taken from Cooper (2008), Morris (2008), Phillips (2008), Soros (2008), Baker (2009), Foster and Magdoff (2009), and Krugman (2009).

1972: Derivatives are invented as investment formulas applied to foreign currency futures (Phillips, 2008: 34); they allow high leverages; currencies begin to be traded in financial markets; in 1973 Black and Scholes published their option-pricing formula based on derivatives (cf. Fox, 2009). The TIE was that investment houses started using derivatives as safe investments. This started a spiral where money involved and leverage used grew and spread around the globe. The crash induced by the failure of Long-term Capital Management in 1998 showed that fail-safe beliefs, approaches and formulas eventually failed.

- Relevant SFTs: *Contagion bursts*: banks substitute for crowded rooms, buses, and planes, thereby spreading contagion more quickly. *Phase transition*: 1st critical value threshold lowered; use of derivatives lowers the 1st critical value threshold such that high-risk and high-leverage investment practices emerge with less initial risk.

1979: The IBM PC triggered the development of computational finance and complex financial-engineering arbitrage instruments. The TIEs were that computers and programming fostered the invention of all sorts of investment formulas; the complexity of formulas and interacting formulas buried in code grew; this made risk-implications more obscure and detection of high risk more difficult. This started a spiral and as computer memory and speed increased, programs and investment methods became even more complex. Supply increased demand and demand increased supply; and traders' use of laptop computers grew worldwide.

- Relevant SFTs: *Irregularity generated gradients*: individually insignificant innovations plus positive feedback fuel rapid, unwatched, and unchecked growth. *Phase transition*: 1st critical value threshold lowered (computers become cheaper; 'quants' move into Finance); emergent degrees of freedom in re-combinations of high-risk tranches are obscured by the complexity of the securitisation packages.

1986: Mortgage-backed securities (MBSs) were invented by UCLA's Professor Richard Roll while he was Director of Mortgage Securities Research at Goldman Sachs from 1985 to *1987*. Tranches were created to mix strong and weak mortgages into loan securitisation packages in order to reduce risk, and incomprehensible computer-based loan securitisation instruments materialised. The TIEs were that the value of houses was treated like the value of cash in a checking account in a bank and that investment banks began to base investment strategies on mortgage-backed assets. This started a spiral where tranche-based, financial-engineering designed securitisation packages took many forms and grew into trillions of US$ in value worldwide. Leverage first increased to 50/1 and when the trend spread world wide, leverage went as high as 100/1.

- Relevant SFTs: *Combination theory*: increased complexity of securitisation packaging resulted in interactive combinations of high-default-risk tranches that multiplied into securitisation packages more likely to show skew distributions and Pareto extremes. *Phase transition*: securitisation packages lowered the risk threshold for taking advantage of mortgage-backs; what emerged was a complex obscuritisation of mortgage-backed risk vulnerabilities.

1999: Glass-Steagall Act was repealed in 1999 by the Clinton administration in the US. The TIEs were that 'deposit-style' banks began taking the kinds of risks associated with investment banks; the US firm Citigroup bought the US firm Smith Barney. This erased the separation between depositor banks and investment banks and all banks were allowed to take highly leveraged high-risk investment actions. This started a spiral where the practice spread to many banks, then to government-based banks (Freddie Mac, Fanny Mae) and US insurance companies such as AIG (others bought small banks so they could begin to pursue risky investments like bigger banks and investment banks), and then worldwide.

- Relevant SFTs: *Contagion bursts*: contagion is bank by bank as opposed to random individuals; speeds up gross risk-taking by banks. *Spontaneous order creation*: banks, as agents, communicate, learn, and influence each other with positive feedback effects.

2000a: $609 billion of foreign reserves in US Treasury securities. By the end of 2004, this had grown to $1.2 trillion (Federal Reserve Bank of San Francisco, 2005) and to $3 trillion by 2008 (US Treasury, 2008). The TIE was that bond interest rates sank, making borrowing cheap; imports cost less than US-produced products. This started the spiral where US Federal Reserve discount rate sank to 0.75 in November 2002; real-estate speculation grew and risk associated with MBSs grew; debt sky-rocketed so that it amounted to ~335 per cent of US GDP by 2006.

- Relevant SFTs: *Spontaneous order creation*: people learned to use cheap money to refinance their houses, often several times; mortgages increased in value, debt-based spending increased, and the fragility of the economy increases.

2000b: Housing bubble starts in US, UK, Spain, and Australia; the US Bush Administration's policy fosters more home ownership, especially among minorities. The TIEs were that the prices of new houses were not bid down but rose as soon as they were put on the market; many houses in Los Angeles (CA, US) were sold with a higher than 'asking price' within an hour; millions of people tried to buy their first house. This started the 'US Bush policy spiral': Get Americans (especially minorities) into their own homes; mortgages were given without checking credit scores or proof of an income; more people bought houses and demand pushed up prices; high prices meant fewer defaults on loans since houses could be unloaded at higher price; less risk led to more home buying; higher prices with less risk of default led to more speculation by real estate brokers, followed by more buying, and so on.

- Relevant SFTs: *Phase transition*: 1st critical value threshold lowered; fostered emergent low-interested based mortgage policies. *Contagion bursts*: banks and lenders stopped asking for credit scores and income statements; borrowers learned about cheap mortgages; towns and builders learned to build housing developments to take advantage; real estate agents started speculative buying and reselling.

2001: 5- and 2-year subprime 'teaser' loans used to buy houses at low interest rates emerged; 1 per cent 'interest-only' mortgages appeared. The TIE was that the US Federal Reserve discount rate was 1.25 in December of 2001 and reached its low of 0.75 in November of 2002; the proportion of subprime mortgages increased and real estate speculation increased. This started the spiral where teaser fixed-rate mortgages increased; US Federal Reserve discount rate increased to 6.25 per cent by July 2006; that is, this meant that when the 5-year fixed-interest teaser loans were converted to variable rates, the latter rose from ~1 per cent to ~6 per cent. Property taxes increased with the value of homes, and there was a vast increase of mortgage-backed investment funds available but risk increased greatly as well.

- Relevant SFTs: *Contagion bursts*: agents connected to banks speeded up the spread of the teaser loan idea. *Preferential attachment*: Some US-based lenders (for example, Countrywide, IndyMac, Washington Mutual, Freddy Mac, Fanny Mae, and so on), became especially well known for offering teaser loans with few questions asked (that is, poor underwriting procedures); they then had links to other banks (often later acquired) such that use of subprime teaser loans spread to most other banks.

2003: Annual volume of securitisation-package issuance reaches $4 trillion as banks leveraged deposits up to a 30/1 to 50/1 and even 100/1 margins based on MBSs; use of securitisation packages spread worldwide by 2005. Each new subprime mortgage was essentially a TIE. This started the spiral where MBSs led to new kind of securitisation package as a TIE; securitised packages increase and leverage increased, the risk of default was increasingly likely, and the economic fragility increased.

- Relevant SFTs: *Contagion bursts* and *Spontaneous order creation*: see above. *Combination theory*: increased complexity of securitisation packaging with increasing leverage results in combinations of skew distributions that compounded into Pareto extremes of vanishing bank liquidity (that is, loss of bank liquidity in amounts never experienced previously in the US or on a worldwide scale).

2006: Bubble in housing prices ended, and house prices began their decline in the US in January. Mortgage defaults sky-rocketed; the 'Minsky Moment' had passed. The TIEs were increasing defaults, more banks at risk, incredibly complicated packages and very high leverage, even worldwide. This started a spiral where subprime teaser loans began to expire and defaults and foreclosures began to rise. The housing market and construction collapsed, mortgage-backed securities became toxic, and liquidity began to collapse.

- Relevant SFTs: *Contagion bursts*: all the connectivities among banks now fostered rapid decline rather than rapid growth. *Preferential attachment* in reverse; that is, the networks and attachments created before 2002–2003 now spread panic and decline.

2006: Five-year teaser loans began to expire thereby forcing owners into variable mortgages which now (in 2006) had higher interest rates. Sub-prime mortgage defaults increased, and the housing bubble expired toward end of year (that is, when houses came onto the market their price was bid down rather than up); the 'Minsky Moment' appeared. The TIE was the beginning of post-teaser defaults; more teasers expired in next two years; the outcome was the 2007 liquidity crisis. This started a spiral where Minsky's phrase *'stability creates instability'* began to apply: housing price bubble led to inflation; increased inflation worries led the US Federal Reserve to raise the discount rate. Mortgage interest rates increased and the spread between teaser rate and post-teaser variable interest rate increased. The foregoing together with increasing property taxes led to an increasing mortgage default rate; this undermined the value of MBS assets and led to economic fragility.

- Relevant SFTs: Same as for 2002–2003 above, except in reverse, that is, *after* the 'Minsky Moment (s)'.

2007: Summer: Subprime crisis hit; home-mortgage defaults and foreclosures sky-rocketed in the US; the collapse of the US firm Bear Stearns began in July; broader bank failures in US in August and the UK's Northern Rock failed in September.

2007: 1 December: Recession began in December (stipulated after-the-fact by the US National Bureau of Economic Research).

2008: September: Crisis-panic hit; US Treasury Secretary Henry Paulson asks for $700 billion bailout; US liquidity crisis spread worldwide; the Great Recession begins.

This account makes clear that there were many instances over the past decades where TIEs spiralled up and created the bubble-build-up phenomena highlighted by the six SFTs. In summary, the build-up to the 2007 liquidity crisis started in 1972 with derivatives being connected to options pricing in the Black-Scholes (1973) options-pricing model. The review of precursors to the 2007 crash also shows various examples of Minsky's positive feedback processes. This justifies our focus on scalability dynamics – that is, positive feedback and additional dynamics that lead to herding behaviour (Hirshleifer and Teoh, 2003) among stock traders. Usually, the emergence of scalability dynamics depends on the presence of tension and connectivity among agents – in this case, stock traders.

While no single scale-free dynamic can be said to be *the* cause of the liquidity crisis, the SFT theory called '*combination theory*' holds that the interaction/multiplication of somewhat skew distributions quite often produces a power-law (Pareto) distribution (Newman, 2005). So how can this help to decide when resilience engineering methods should be used? We can ignore the many examples of short-lived market bubbles that relatively soon return to random-walk efficient-market trading without intervention. But we must be able to recognise the bubble build-up that can reach a crash point – what we will term the Critical Point in Chapter 9.

Chapter 6
Taming Manias: On the Origins, Inevitability, Prediction and Regulation of Bubbles and Crashes

Jeff Satinover and Didier Sornette

Can one desire too much a good thing?
Don Quixote, I:I:IV

We consider the recent financial crisis as an overlapping sequence of interdependent financial bubbles followed by their collapse. Governments and regulatory agencies have made it a prime goal to moderate future crises. Many attempts at financial, economic and social engineering are plagued by an illusion of control typical of complex systems for which we offer some suggestive mathematical models. Furthermore, control may not only yield no benefit, but at times may exact perverse new costs. We argue that markets and economies in general are truly 'complex systems' in a technical sense. As such, they are intrinsically characterised by periods of extremity and by abrupt state-transition and spend much time in a largely unpredictable state. Yet they sometimes enter periods of pre-crisis when they are predictable. In consequence of this we argue that the most extreme events – the most influential ones – are susceptible to (probabilistic) prediction. In light of this analysis, we offer a small number of perhaps counter-intuitive suggestions, for example, that many of the present interventions in the 'liquidity crisis' are ill-advised and possibly dangerous – for example, the widespread attempts to artificially stimulate consumption in the absence of precautionary reserves and in the presence of huge liabilities; as an example of real-world, large-scale resilience engineering we suggest that bubble-prediction should be a central part of financial regulation.

As this chapter is being finalised (December 2010), the world has gone through the worst financial crisis since the Great Depression. It is not yet clear whether we have been through the worst or if worse is yet to come. In many respects, economies have stabilised and many of the largest have emerged from technical recession. For self-evident reasons, governments are touting the recovery, the severity of job-losses notwithstanding. But a larger-scale view is not so rosy. There are unfunded liabilities in many major countries on a scale from which no national economy has ever previously escaped without either inflating its currency to a devastating extent, or via war.

The present financial state of the world follows the collapse of five sequential, partially superimposed, and interdependent bubbles (Sornette and Woodard, 2010). A so-called 'new economy' Internet, communications and technology bubble that began in the mid 1990s, and which burst in the crash of mid 2000. This crash was moderated by a liquidity-to-real-estate bubble deliberately furthered by US Federal Reserve monetary policy. Following the technology crash, the US Federal Reserve lowered the Federal Fund Rate from 6.5 per cent in 2000 and to 1 per cent in 2003–2004. This extraordinary access to cheap money helped fuel a real-estate craze, via the wild expansion of mortgages it incentivised a period of intense creativity in financial-instrument derivatives. These instruments were believed to exploit the real-estate market more efficiently (Mortgage Backed Securities) and it was also argued that the excessive (or unrealistic) valuations in real-estate asset value could be associated with *lower* rather than *higher* levels of risk (Collateralised Debt Obligations, designed to separate the risk of bundled assets into uncorrelated tranches). As financial engineering grew ever more sophisticated and complex, in the presence of tremendous liquidity, commodity bubbles began inflating as money sought high but stable rates of return in ever more markets. The illusory wealth thus accumulated helped inflate a stock market bubble that peaked in 2007. The technology bubble is notable for the degree to which it was based on, and helped foster further, innovation in telecommunications and computation. Even if a new economy was not actually created during this bubble, a genuinely new and extraordinarily valuable set of industries and products did emerge. But the remaining four bubbles arguably lack any true 'product', they represent the internal recycling and revaluation – in albeit highly creative fashion – of existing 'real' products. (New housing stock was built in the United States. It remains to be seen how much will last and how much innovation it generated.)

Financial innovation and financial markets have always been an essential part of economic growth in both developed and developing countries. Financial markets not only serve as critical indicators for investors, but as liquidity providers. The price of a stock, and especially changes in its price, represents a quantified assessment of how valuable a company's products or services are. In general, a healthy stock price, or price rise, increase the likelihood that a loan for a new product will be repaid. One characteristic of a bubble, which can only be seen a posteriori, is that an increase in price begins to magnify this likelihood. However, the importance of finance can also be exaggerated. Is it a good thing if the corporate strategy of a firm becomes driven largely by its stock's value, or if the personal wealth of corporate decision-makers derives largely from their stock-options (Broekstra et al., 2005)?

Inter alia, financial markets play a role as information providers – they provide (at least one) measure of the value of an asset class (and of the underlying assets). But decision-making within a corporation can become too dependent on purely financial incentives. If, for example, an executive's bonus is tied 'excessively' to the company's stock price, managerial actions may (deliberately or not) have the primary goal of further increasing that price, irrespective of other perhaps more fundamental measures of corporate health. This kind of action enhances the risk of a self-perpetuating and ever-inflating bubble, and also the likelihood of a destructive or exaggerated correction; estimated numbers in 2007 dollars were $250 billion, $4,700 billion and $26,400 billion respectively. The situation is especially dangerous if widely disparate financial markets and economies are critically interlinked as they are today because of very sophisticated financial instruments (as well as general interconnectivity via communications technology). Blanchard (2008) offered a very informative snapshot of how what happened as the subprime mortgage crisis in the US cascaded into a worldwide crisis: if estimated subprime losses at October 2007 were defined as equalling -1 unit, then the cumulative worldwide loss in Gross Domestic Product equalled ~-20 units and the decline in worldwide stock market capitalisation equalled ~-100 units. The estimated numbers in 2007 dollars were $250 billion, $4,700 billion and $26,400 billion respectively.

Risk Diversification, Correlation and Unanticipated Disaster

The world of financial engineering has seen extraordinary growth in recent years and many extremely sophisticated methods have been developed to assess and distribute risk, largely by the development of new instruments derived from packaging and then re-dividing large number of underlying simpler instruments. While many of these methods are quite sophisticated, they generally fail to incorporate the crucial insight that financial markets and economies have become very complex systems. Financial instruments are designed making a convenient and desirable assumption of independence. Complex systems typically contain many instances of hidden interdependences, tight couplings and other subtle (and inconvenient!) features. In the following, we list some of the most important (see also Chapters 2 and 4):

- Complex systems are usually open and dynamic – the underlying components of the system are in flux. Nonetheless, complex systems usually demonstrate stability of patterning which lends itself to a mistaken presumption of equilibrium, as in classical economics and control theory.
- Most frequently the stability of patterning may be considered a 'meta-stability'– it includes multiple quasi-stable states with dynamic, abrupt and difficult to predict transitions among states.
- Complex systems often have a memory. The future state depends not only one or more preceding states but upon

the dynamic sequence of preceding states – that is, they demonstrate path dependency. This feature lends these systems the power of self-organisation.

- Complex systems often consist of nested hierarchies of smaller-scale complex systems. This is most evident in the neurobiology of the brain where, for example, cortical brain tissue forms a self-organising complex systems 'sheet', but is itself at a lower-level composed of cortical processing units which 'compute' an output passed up to the higher level. The cortical units are themselves composed of individual units (neurons) whose computational capacity arises from the complex systems nature of their internal components. In the other direction, economies and markets may be thought as a composed of many individual brains, or agents (Satinover, 2002).
- Complex systems often yield outputs that are emergent: the interactions among agents/individual units may be deterministic, but the global behaviour of the system as a whole conforms to rules that are only rarely deducible from knowledge of the interactions and topology of the system. As mentioned in Chapter 2, Financial Services systems are intractable, which means that it is impossible fully to specify them.
- In complex systems, the relationship between input and output is typically non-linear so that a small perturbation may yield a very large overall disturbance while a large perturbation may be absorbed with little or no effect. Complex systems are typically exquisitely sensitive and at the same time resilient, in ways that are difficult to predict.
- Complex systems are also characterised by having both negative (damping) and positive (amplifying) feedback loops. The output of the system alters the nature of the (next) input (cf. Chapter 3).

As we proceed through the chapter, we will see most of the above features in play in the systems we present. All of these features make optimisation and engineering resilience very challenging – especially when the systems of interest are as large and interdependent as international finance and their associated economies.

Unanticipated Coupling as a Basis for Failed Risk-Management

We begin with a simple but incisive example that illustrates much of what went wrong in the global Financial Services system from a quantitative risk-management perspective, that is, where attempts at optimisation (risk minimisation) generated perverse results, contributing to the initial 'fracture' that resulted in the above described loss of wealth. The essential problem was a failure to anticipate correlation among default risks and its impact on the global Financial Services system.

Consider this anonymous quotation from 'Confessions of a Risk Manager' in *The Economist*, 7 August 2008:

> Like most banks we owned a portfolio of different tranches of collateralised-debt obligations (CDOs) ... Our business and risk strategy was to buy pools of assets, mainly bonds; warehouse them on our own balance-sheet and structure them into CDOs; and finally distribute them to end investors. *We were most eager to sell the non-investment-grade tranches, and our risk approvals were conditional on reducing these to zero.* We would allow positions of the top-rated AAA and super-senior (even better than AAA) tranches to be held on our own balance-sheet as the default risk was deemed to be well protected by all the lower tranches, which would have to absorb any prior losses. [Italics added]

Consider a portfolio of two wholly identical Debt Obligations (DO) with the same face value ($1,000) and default risk (10 per cent). On the assumption that the default risks are uncorrelated (independent), we may compute the risk-adjusted associated value of this portfolio as 2 × $900 = $1,800. We now re-securitise this portfolio into a new portfolio of two 'synthetic' IOUs – both of face value $1,000, by contractually re-assigning the rewards associated with the three different outcomes (no default, one default, two defaults). The key step in creating these new securities (which we will call MBSs, to reflect the real-world collateralisation of these instruments by real-estate mortgages) is that one instrument has a much lower risk of default and the other a much higher risk of default, than either of the two underlying real securities. In this way, the risk management mandate cited above may be readily fulfilled, that is, to sell off high-risk debt.

The new, synthetic MBSs are defined as follows. The low-risk, 'senior tranche' MBS is a debt instrument with face value $1,000. If neither of the underlying securities defaults (they together repay the full $2,000), the senior tranche is worth $1,000. This will happen with probability 81 per cent. If either of the underlying securities default but nevertheless repay the full $1,000, the senior tranche is still worth $1,000 (by agreement). This will happen with probability 18 per cent. If both securities default, the senior tranche defaults as well and is worth $0. This happens with probability 1 per cent).

The high-risk, 'junior tranche' MBS also has a face value of $1,000 but is structured differently. Like the senior tranche, if neither of the underlying securities defaults, the junior tranche is worth $1,000. However, if either (or both) the underlying securities default, the junior tranche is worth $0.

In short, the senior tranche has a 1 per cent probability of defaulting while the junior tranche has 1 per cent + 18 per cent = 19 per cent probability of defaulting. Both have the same face value, but almost all the risks are now associated with the junior tranche and almost none with the senior tranche. The junior tranche is the synthetic equivalent of a 'non-investment grade' loan. The price for the senior tranche is 99 per cent × $1,000 + 1 per cent × $0 = $990 and for the junior tranche 81 per cent × $1,000 + 19 per cent × $0 = $810).

It is important to note that this analysis assumes independence of the defaults. If instead the defaults are correlated, the reallocation of risk is an illusion. If the correlation is perfect, the synthetic debt instruments collapse into ones with risks identical to the underlying real ones but now very badly priced. As both underlying securities either do or do not default in tandem, there is no state where the portfolio is worth $1,000. There is a 90 per cent probability that is worth $2,000 and a 10 per cent probability that it is worth $0. Both the senior and junior tranches are now worth $900 and each has a 10 per cent risk of being worthless. The 'Investment Grade' synthetic debt instruments, for which a very high premium was paid and which the bank 'hoarded', turns out to be a hidden but highly explosive cache.

The problem of correlation among assets is a subtle problem. Mathematical tractability assumes independence across assets, across asset classes and over time in price time series. Both the efficient market hypothesis (Fama, 1970) and the assumption that prices follow a random walk have sufficient evidence to make them plausible, at least in weak form, and they are both convenient. But plausibility, a certain amount of evidence, and mathematical tractability do not mean that these hypotheses are true for all markets. Furthermore, as we will see, departures from them are often the most influential events in markets, for better or worse. The same underestimation of correlations at times of crisis was also the nemesis of the famous hedge fund Long-term Capital Management (MacKenzie, 2006).

Cooperativeness in Markets as a Mechanism for Dependence

The example in the previous section illustrates how the assumption of independence led to a disastrous misvaluation in the securitised mortgage market. In the first section of this chapter, we briefly described how this disaster in one domain combined with the tight coupling in Financial Services systems led to a worldwide avalanche of vastly greater scale.

In complex systems, global non-linear patterning arising from local behaviour is the rule rather than the exception. Small changes in temporally or spatially localised domains may lead to unexpectedly large global effects rather than being averaged out. In natural and human systems, complexity is the norm and analytical tractability the exception. The so-called Minority Game (MG) is a very simple model of a complex system, namely a market of a scarce resource. It illustrates how cooperativeness may arise in distributed systems even if no proper cooperation takes place among the agents who collectively make up the system (Challet et al., 2005). By agents, we specifically mean software objects that interact with one another and/or the system as a whole, according to a set of embedded rules in (usually) uniform, discrete time-steps. Agent-based systems may be structured so that their collective output (perhaps a simple addition of individual outputs) forms a simple, non-complex system. However, the most interesting agent-based models generate outputs that represent complexity, as defined previously.

The assumption that asset values and risks are independent is often supported by examining the chains of observable causality that may exist among networked agents. (In a designed model, these rules of mutual influence are specified in advance and not merely known by observation or guesswork.) If such evident linkages are few (whether direct or indirect), independence is presumed. In the MG, no such linkages exist. Agents decide how to act based on internal computations in response to a commonly known history of market valuations; and their strategies are allocated randomly from a set of all possible strategies. Yet under certain circumstances, the system of agents seems to conspire to behave cooperatively as reflected by price changes in the market.

The Minority Game (MG) as a Model Market

The Minority Game (MG) is a discrete-time, multi-agent model for a market in a scarce resource. The time-series output of the game shows many features in common with demand-imbalance (price) in real markets, for example volatility clustering. Agents cast a binary vote (for example, up or down) at each time-step. Agents win one point if they cast their vote with the minority; they lose one point if they follow the majority. Agents choose whether to vote up or down according to a strategy table that maps an immediately-prior fixed-length history ($\mu(t)$, of length m) of the winning overall state to a vote choice. For example, a given strategy table might indicate for $m = 3$, 'up, up, down', vote 'up', Every agent randomly selects a given number of such strategy tables when a game begins. At each step, it deploys that strategy table which would have accumulated the most number of (hypothetical) points had it always been used, given the real history of the game. Exact details may be found in Challet and Zhang (1997) and in Satinover and Sornette (2007).

The sequence of minority state values forms a time-series with analogies to a time-series of market returns. An important characteristic in both is the volatility (dispersion) σ. We know that in markets, in general, volatility of returns is not randomly distributed over time, but that it clusters. In the Minority Game (MG), volatility σ varies in a precise way with m. This variation shows a remarkable and subtle structure: as m increases from $m = 2$, σ decreases sharply until at a critical value of $m = m_c$ (given N, the number of agents), σ again increases. In other words, σ demonstrates a phase transition (another hallmark of a complex system).

There are other important features of how σ varies with m. Consider that, if all agents *always* deployed strategies at random; or if all agents simply voted at random with equal probability, σ would remain at a value easily computable in advance as a function of N, and would remain constant regardless of m. Label this value σ_{exp}. At minimal values of $m < m_c$, σ is unexpectedly high – $\sigma > \sigma_{exp}$. More remarkable is the fact that at m_c, where σ attains its minimum, this minimum is *below* σ_{exp}. In other words, at a critical memory length, agents appear to be acting as though allocating resources cooperatively, attaining a level of efficiency and 'sharing' that exceeds what can be expected from system of agents selecting their actions randomly. For values of $m > m_c$, σ increases again, but approaches σ_{exp} only asymptotically.

Even more illuminating from the perspective of unanticipated cooperativity is the fact that this same phase-transition structure remains close to unchanged *if all actual history sequences $\mu(t)$ are replaced by randomly generated sequences* (Cavagna, 1999). In other words, this kind of cooperativity does not even necessarily reflect common knowledge of the prior history of the system – it requires only that all agents act on a common information (or misinformation) set. We may see here an analogue to how the role of financial markets as information-providers (a function of Financial Services systems as suggested by Merton, 1995) can become exaggerated.

Figure 6.1 illustrates these phenomena: in this deceptively simple model, we see that a mere shared misperceptions (random histories), may have significant impact on the degree of correlation among actions. This is the case even when agents are disparate, base their actions on different rules, share no information and whose individual actions are apparently independent of the actions of any other agents.

Most models of complex systems are analytically insoluble. But in its simplest variations, the MG is solvable and there is therefore no mystery whatsoever in understanding how its collective cooperativity arises. But for systems beyond analyticity (and even small changes in the MG rules will yield little difference in the phase transition curve but no analytic solutions), this cooperativity would remain unfathomable, absent knowledge of the MG proper as we've just described it. For real-world financial

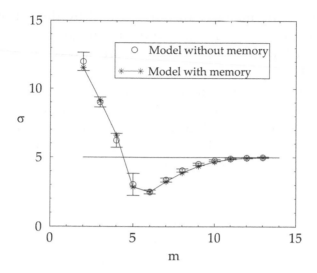

Figure 6.1   Variation in volatility in the MG

Source: Cavagna, 1999.

engineering, the bottom line is: correlation emerges unexpectedly and as a result any analysis which assumes independence must be clearly identified. For example, projections of added value when creating MBS tranches should be recalculated using the inverse assumption, that is, perfect correlation. Investments whose projected risk increases significantly when correlation is assumed, should be viewed as significantly riskier – although to an unknown degree – than under the presumption of no or even less correlation. This recommendation of course leaves open the question of the likelihood of each possible degree of correlation when we are dealing with 'known unknowns'.

Illusion of Control

Prediction in complex systems is characterised by another factor that may be a challenge to successful resilience. A widely shared 'correct' prediction leads to generally common action that may alter the predicted results and thereby make such 'correct' predictions wrong. Stated boldly, the 'correct' predictions in such a case are by definition incorrect. The MG offers a concrete quantitative example of this dilemma.

In the MG, the majority of players will by definition always lose. These losses accumulate and are on average distributed among the players. However, there are always a few players who in the long run win. Or is it possible for players to devise a strategy that will win predictably? One case is if certain privileged players know in advance the distribution of strategies deployed by the rest of the market. But in this case a known, asymmetrical advantage is introduced – akin to insider trading in a real market. Another case, which does not introduce an asymmetrical advantage, is if a small number of agents deliberately deploy that strategy which in the agents' internal performance calculations has performed worst. On average, agents that use this 'perverse' rule not only outperform standard agents, they actually generate a positive return (Satinover and Sornette, 2007). The illusion of control is exemplified by agents who play randomly without any optimisation and thereby outperform their optimising siblings. The underlying driver for this paradoxical result is the frustration introduced in the minority payoff of the game: because a winning strategy gets progressively adopted by the majority, it is bound to fail. Backwards optimising agents are therefore always 'fighting the last war' and cannot react fast enough to face this self-generated, ever-changing market structure. Completely random strategies are better adapted to such a world.

The illusion of control is directly related to the fact that agents' 'correct' calculations consist in an assessment of how a strategy would have performed, that is, in the past – but without taking into account how the game conditions would be altered if every agent did the same. In the MG, agents cannot calculate the impact on those game conditions (the state of the 'market'), if they had deployed a different strategy. An agent may only consider what actually happened in the past. It is possible to grant every agent a probabilistic ability to assess the impact of its choice of strategy. But as soon as one grants every player the ability to do this, the game no longer reflects the behaviour of true markets and settles instead into a stable Nash equilibrium – a system that is no longer dynamic. Were one to grant only some agents the ability to do this, asymmetry is once again introduced and these superior agents will unsurprisingly outperform their 'less intelligent' fellows.

Dragon Kings

One of the challenges facing policy-makers and risk-managers is that exceptional and unimaginable events by definition are much less common than non-exceptional ones, and therefore may not lend themselves to meaningful statistical analysis and prediction. It has become second nature throughout the natural and social sciences to treat such events as 'outliers' – not necessarily because they are true 'exceptions' to an accurately denoted rule, but simply because they seem to prevent the development of such a rule when included in the dataset. Examples gathered from a single domain are generally so sparse as to preclude generalisation. However, if extreme events from seemingly unrelated domains plausibly can be understood as arising from a common underlying mechanism, it would be possible to aggregate data from these domains. From this aggregation and commonality of cause, meaningful intra-domain understandings and predictions may be proposed.

As developed in Sornette (2009), the concept of extreme meaningful (in contrast to meaningless!) Dragon King's outliers can be applied to a many different systems in a broad range of conditions (for example, distribution of city sizes, distribution of velocity increments in hydrodynamic turbulence, distribution of financial draw-downs, distribution of the energies of epileptic seizures in humans, and distribution of earthquake energies). These Dragon Kings may coexist with power laws in the distributions of event sizes, and suggest mechanisms of self-organisation otherwise not apparent in the distribution of their smaller siblings. Dragon Kings are commonly associated with a phase transition. If a phase transition can be detected as it happens, it may be seen as an abrupt increase in the probability, or risk, of an extreme event. Practical examples include ruptures in materials and rupture in financial bubbles, which are briefly reviewed below. From a resilience engineering perspective, the ability to detect Dragon Kings would increase resilience by improving a system's ability to anticipate and to respond.

Figure 6.2 illustrates the concept of Dragon Kings in the financial sector. It shows the non-borrowed reserves of depository institutions as compiled by the Federal Reserve of St Louis from the late 1950s to February 2010. This time series exhibits

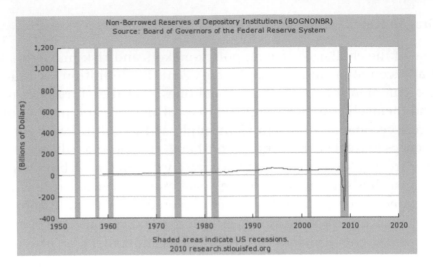

Figure 6.2   Illustration of the concept of Dragon Kings in the financial sector

Source: Reproduced from the Federal Reserve of St Louis. From: http://research. stlouisfed.org/fred2/series/BOGNONBR. Monthly data up to February 2010.

fluctuations in the range of tens of billions over its first 50 years, which are barely visible on this graph where the vertical axis covers a range from -400 to +1,200 billions of dollars! Then, two Dragon Kings occurred in short order, the collapse to close to -400 billion in 2008, followed by an extraordinary hoarding of more than one trillion of dollars as measured in February 2010 and growing. The two last Dragon Kings are complete outliers to the fluctuations in the previous 50 years. Indeed, 50 years of history of the evolution and variations of the non-borrowed reserves of depository institutions tell us absolutely nothing about the recent extraordinary events.

Black Swans and Dragon Kings

Although the term Black Swans (Taleb, 2007) has entered popular parlance as a generic label for rare, extreme events, it has a very specific meaning and underlying theory. It represents an advance in our understanding of how event sizes are distributed, from a simple ('bell-shaped') Gaussian distribution of independent events to a more sophisticated power-law distribution ('fat tails' and sharp central peak). Power-law distributions are typically found across social and natural domains where events are the output of a complex system. Relative to this, Black Swans are events in the fat tails of the distribution. They are therefore more common than events in the comparable region of a Gaussian distribution, but are also completely unpredictable, just as distributions that arise from completely independent events (a non-complex system). In this view, a massive earthquake is one that started small and did not stop. The underlying generating mechanism is the same for large and small earthquakes. The stopping times of their self-generating growth processes are unpredictable, the only predictable property being that longer stopping times are less common (according to a power law) than shorter stopping times. The stopping times are unpredictable because there is nothing that a priori differentiates an earthquake that grows to scale s and stops from an earthquake that grows to scale s and keeps growing.

However, complex systems are not only characterised by a power-law distribution in event sizes. They also typically represent abrupt shifts in the entire regime (rules) by which the underlying interacting elements are self-organised. They are empirically characterised by very extreme events above a certain size that clearly do not fit a power law. The abrupt change from a power-law fit to a non-power-law fit strongly suggests a regime change or phase transition. These very extreme events are rarer than merely extreme events. As argued in Sornette (2003a; 2009), the abrupt change to a non-power-law distribution strongly suggests a different generating mechanism. If so, there may exist signatures in the growth period of these Dragon Kings that differentiate them from events that may or may not become 'black swans'. The presence of such signatures may represent an elevated level of risk that can be quantified.

Paris: The Dragon King of French Cities

In many countries, including France, the distribution of city population sizes is a very good fit to a power law with exponent ~1 (Gabaix, 1999). This scaling follows from the law of proportional growth, also-called Gibrat's law (Malevergne et al., 2008; Saichev et al., 2009). However, as shown in Figure 6.3, the population of Paris is several times larger than expected from the distribution of the other cities (Laherrère and Sornette, 1999). Malevergne et al. (2005) have shown that power-law distributions are degenerate members of a large family of stretched exponential distributions, that is, $\sim exp[-(S/S_0)^c]$, where S is the city size, S_0 is a characteristic city size and c is an exponent (a value c smaller than 1 defines the stretched exponential family). Because the best empirical stretched exponent $c = 0.18$ is small, the straight line in Figure 6.3 qualifies a power-law distribution, with an exponent which turns out to be close to 1 (Zipf's law).

However, the most interesting feature of Figure 6.3 is the remarkable deviation of Paris: rank 1 (corresponding to left-most value log(1)=0 in the abscissa presented in log-scale), the largest city of France. Clearly, this city does not conform to Zipf's power law. Other major centres of civilisation demonstrate a similar departure from the otherwise remarkable performance of Zipf's law, for example, London in the UK Should these major centres

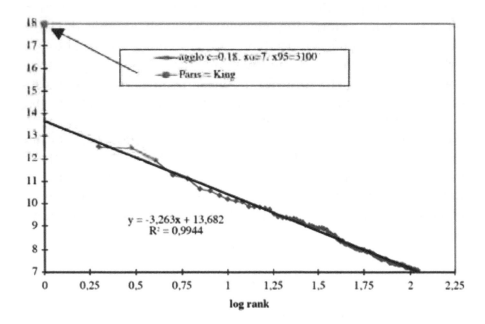

Figure 6.3   Rank-ordering plot of the population of French cities

Note: The arrow shows Paris.

Source: Laherrère and Sornette, 1999.

be considered 'statistical outliers', or would it make more sense to treat them as enormously important instances that, although rare, in some sense are Dragon Kings that exhibit a dynamic distinctly different from their normal and Black-Swan-sized fellows. In the case of the great cities, the idea of making sense of a civilisation by throwing out only their capitals highlights the importance of trying to understand the common uniqueness of Dragon Kings across domains. We hypothesise that city Dragon Kings arise because of internal *positive* feedbacks among growing elements within a city. A common example of this, well-known and exploited by savvy real-estate and business developers, is how a neighbourhood with many competing businesses *of the same kind* generates greater growth for all. Thus where one finds newly opened MacDonald's, one is also very likely soon after to find a Burger King built right across the street, or even next door. The naïve outsider is likely to think that it would be best to build the Burger King in a location with less competition. This is true in the seeding of new potential locations, but not at all true when a region shows signs of growth. As a consequence, the rate of growth can accelerate, giving rise to a 'super-exponential' growth regime, leading to the emergence of the Dragon King.

Self-Organised Criticality and Dragon Kings

Self-organised criticality has been proposed as a general phenomenon, occurring in many natural and social systems. Open systems that are slowly destabilised often tend to self-organise their dynamics at a 'critical point' at the border between order and chaos, characterised by the spontaneous generation of fluctuations in the form of 'avalanches' or cascades of instabilities spanning many orders of magnitude (Jensen, 1998; Sornette, 2006). Sandpile models that slowly are destabilised (cf. Chapter 5) often readjust via a hierarchy of avalanche sizes. These have been proposed as an example of models of such generic behaviour of complex systems, including the dynamics of supply chain networks (Scheinkman and Woodford, 1994), economies of firms (Andergassen et al., 2004) and the cascades of bank failures. One signature of self-organised criticality is that avalanche sizes show a power-law distribution. This has led to the claim that large avalanches, such as great earthquakes, are intrinsically unpredictable as mentioned above (Geller et al., 1997).

Here, we present a system analogous to a sandpile driven slowly out of equilibrium, which readjusts via a hierarchy of avalanche sizes, whose distribution has a power law ('fat tail') regime coexisting with Dragon Kings. The importance of this example lies in particular in the simplicity of its ingredients, which makes possible a full theoretical understanding of its detailed properties. This example studied by Jogi and Sornette (1998) has not been previously emphasised in the present context to illustrate the coexistence of Dragon Kings and power laws.

Consider a two-dimensional lattice in which each grid point is assigned an independent random number drawn from some distribution with unit mean and variance (here an exponential

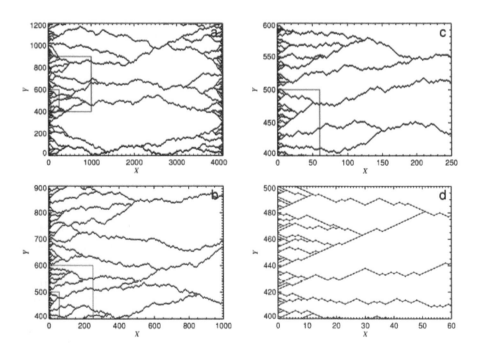

Figure 6.4   **Typical set of optimal configurations for optimal paths of length W=4,096 and for 0≤y≤1,200**

Note: (a) global system [grey-framed boxes outline regions of succeeding plots]; (b) magnification of the largest box in (a); (c) magnification of the largest box in (b); (d) magnification of the box in (c).

Source: Reproduced from Jogi et al., 1998.

distribution with unit mean and variance). The random number carried by each grid point can be interpreted as a local adsorption energy (in physical chemistry), or a local toll cost (in supply chain economics) or pinning strength (in physics), depending on the application. One then asks what is the static optimal ('zero-temperature' in the jargon of physics) configuration of a directed path running along this lattice with its two ends fixed at the two opposite boundaries. Such path configurations minimise the total sum of the random numbers visited by the path on the lattice. In the physical chemistry context, this corresponds to finding the configuration of the adsorbed polymer with minimal energy. In the supply chain context, one is interested in finding the directed trajectory from one fixed departure point on the left of the lattice to the arrival point on the opposite right boundary that minimises the sum of toll costs accumulated by crossing all the grid points along that trajectory. The solution of the global optimisation problem is known in the mathematical physics literature at the 'random directed polymer (RDP) at zero temperature' (Halpin-Healy and Zhang, 1995). Figure 6.4 shows four levels of magnification of the set of optimal paths that minimise the sum of the random numbers of the grid along their lengths, in a lattice of 4,096 × 1,200 grid points. The paths shown are the optimal paths which maximise the sum of the random numbers along their lengths, with fixed ends at the same altitude y, where y scans the interval $0 \leq y \leq 1,200$. This global optimisation problem to determine the minimal paths, which appears at first view as computationally complex, because non-local (the best path globally is not obtained by gluing local best paths), can be reduced by a transfer matrix approach to just polynomial complexity (and not NP-completeness) (Kardar and Zhang, 1987).

As depicted in Figure 6.5, let us now define an avalanche by the area S spanned by the transition from the optimal configuration at the ordinate position y of the two anchoring points in the left and right boundaries to the next position y+1, that is, S is the area interior to the perimeter formed by the union of the two optimal RDP configurations at y and y+1 and the two vertical segments $[(0,y);(0,y+1)]$ and $[(W,y);(W,y+1)]$, where W is the width of the lattice (W=4,096 in the example shown in Figure 6.4). An avalanche is thus the area swept by the optimal

Figure 6.5   **Schematic representation of 'avalanches' between successive optimal RDP paths fixed at their two end points**

Note: The successive avalanches are represented in different grey scales.

Source: Reproduced from Jogi et al., 1998.

path when driven incrementally by its two extremities. An avalanche thus corresponds to the jump in optimal configuration paths when moving slowly the two extremities of the path, grid point by grid point.

Theoretical arguments based on the hierarchical structure of quasi-degenerate RDP and numerical simulations presented in Figure 6.6 show that the distribution of moderately sized avalanches is found to be a power law $P(S)dS \sim S^{-(1+m)} dS$ with exponent $m=2/5$. This power law is valid up to a maximum size $S_{4/3} \sim W^{4/3}$. This power law regime is qualified by the straight part of the curves in Figure 6.6, There is another population of avalanches that, for characteristic sizes beyond $S_{5/3}$, are distributed with $P(S) dS \sim \exp[-2(S/S_{5/3})^3]dS$, also confirmed numerically. This second population is at the origin of the shoulder and the fast drop-off for $S/W^{5/3} \sim 1$. This exemplifies the coexistence of (i) a power-law

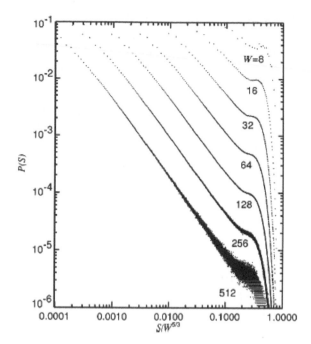

Figure 6.6   Probability density function P(S) of avalanche sizes S as a function of the rescaled variable S/W5/3 for lattice width W varying from 8 to 512 on a log-log plot

Source: Reproduced from Jogi et al., 1998.

distribution of avalanches up to a maximum size $S_{4/3} \sim W^{4/3}$ and (ii) of characteristic Dragon Kings with typical sizes $S \sim W^{5/3}$. The first population corresponds to almost degenerate equivalent optimal paths, providing a direct evidence of 'weak ergodicity symmetry breaking' in random media, and defines the 'self-organised' power-law avalanche regime. The second population is associated with different optimal states separated by the typical fluctuation $W^{2/3}$, which is characteristic of a single isolated RDP (or optimal path) of length W.

A more pointed and quantified illustration of Dragon Kings in finance is provided by a careful analysis of market 'crashes' and the role of positive feedback in generating them (cf. Chapter 3).

Crashes: The Dragon Kings of the Market

The Black Swan theory of how market events scale with frequency is best captured in snapshot by simply examining their actual distribution. If we do this, we find three things. First, the frequency of events by size (market losses) generally declines according to a power law. Second, the most extreme events (which like Paris in France are scarcely to be rejected) occur more frequently than projected by the power law – and they are also the events with the longest and broadest socio-economic effects. Third, events that do fall within the power-law regime may be adequately modelled as independent. The modelling of the most extreme events may, however, require that the idea that they occur at random is rejected – they are strongly coupled to precursory patterns exhibited by financial markets.

We argue that the two processes differ in terms of dependence. That is, a relatively abrupt transition occurs between a market in which successive unit-time losses (or gains) are independent of each other and a market in which such losses (or gains) develop long-range positive dependence. There is a significant degree of positive feedback between the size and sign of successive price changes in the latter process but not in the former. Consider a hypothetical crash-sized 30 per cent draw-down that occurs as three successive days of 10 per cent drops. A single one-day 10 per cent market drop appears in the NASDAQ composite index on average once every four years, or 1,000 trading days and can thus be said to have a frequency or probability of 10–3. Three successive such events naïvely appears to have a probability of occurring of 10–9, that is, once every four million years. But successive crashes of such sizes or larger have occurred in markets all over the world during the last few decades.

The flaw in our calculation is the same as in the creation of differentiated MBS risk tranches, namely independence. It should be intuitively clear that once a market has dropped enough to induce panic, the price-setting mechanism becomes much more highly influenced by sheer herding than following series of small changes or fluctuations. In other words, the correlation between successive events increases. The risk of a series of such events is accordingly higher. As shown in Figure 6.7, the empirical distribution of draw-downs suggests two relatively distinct regimes, which means that events above a certain size represent a change in regime (or phase). This view does not require the sharp discontinuity between regimes near the transition point that can be found in many physical systems, such as the optimal path example of Jogi and Sornette (1998) and in simple market models such as the MG. We expect, rather, that the regimes interlace to a degree near a transition point. The important point is to understand that the *risk* of a Dragon-King-type event rises rapidly from near zero to something very significant near this transition point – especially as these events are so dangerous. The ability of a Financial Services system to detect Dragon Kings will therefore influence its ability to withstand the impact of negative events associated with bubbles and market crashes.

Indeed, it would be a serious mistake to neglect, or ignore, the relationship among Dragon Kings and the pre-existing regimes from which they emerge. The concept of phase transition and the various mathematical techniques that express and quantify, point toward a unified conceptual framework with a potential for prediction of crises (Sornette, 2002).

Johansen and Sornette (2006) found that two-thirds of the Dragon Kings identified in the distribution of draw-downs in exchange markets (US dollar against the Deutschmark and against the Yen), major world stock markets, and US and Japanese bond and commodity markets were crashes that followed large bubbles. Crashes in this light may be understood as the way that pre-existing bubbles deflate (Sornette, 2003a). The bubbles themselves (massive 'draw-ups') are similarly associated with various positive feedback mechanisms that generate a faster-than-exponential regime of growth that is ultimately unsustainable (Jiang et al., 2010). These positive feedbacks are mediated by, inter alia, portfolio insurance trading, option hedging, momentum investment and imitation-based herding – all applied with an inadequate appreciation of growing dependencies across geography, asset classes and in price changes.

A critical question for investors, risk managers, regulators and policy-makers is therefore whether this change of regime can be detected as it is occurring. A second question is what actions ought be taken if such a regime change is detected – that

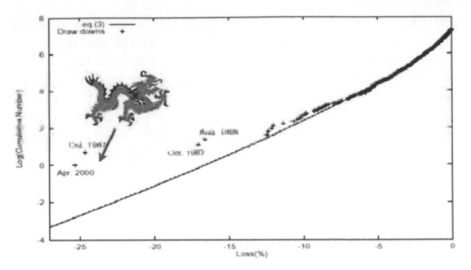

Figure 6.7   Draw-downs (+) in the NASDAQ Composite index

Note: All points that fall significantly off this line turn out to be associated with major crashes.

Source: Johansen and Sornette, 2001.

is, is it possible to mitigate at least some of the most extreme consequences of an abruptly collapsing bubble? From a resilience engineering perspective, the possibility of detecting a change in regime offers at least an opportunity for intervention. It would be an excellent example of how resilience engineering attempts not be enslaved to the past.

Detection and Prediction of Crashes in Markets: An Example

From 1995 to the present, Sornette and collaborators have developed a formalised set of procedures to detect and quantify the risk of an asset market bubble and collapse, based on a *log-periodic (nonlinear oscillatory) power law* (LPPL) signature (Sornette, 1998) in the price series. Inter alia, they have successfully predicted the timing of the recent collapse in oil prices and the US housing market. Most recently, July 2009, they published their prediction of a collapse in the Shanghai Stock Index (Bastiaensen et al., 2009; Jiang et al., 2010). In August, the index began its decline, falling 20 per cent in the two weeks after its peak. Their prediction is summarised in Figure 6.8.

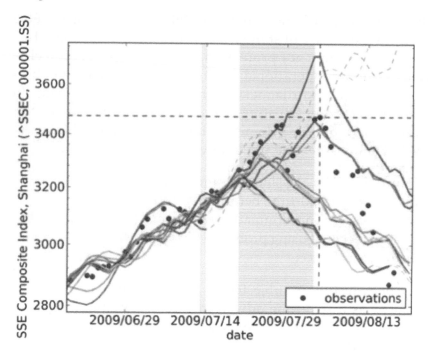

Figure 6.8   Ensemble of most-likely LPPL fits for a collapse at the 80 per cent confidence level

Note: The actual drop began in the first week of August 2009.

Source: Jiang et al., 2010.

In earlier work, Sornette and collaborators made both a priori and (controlled, that is, back-tested) *a posteriori* predictions on tens of markets showing a LPPL signature of a bubble. Their record has been remarkably good, but with a common criticism that they predicted 'nine out of five' collapses. The criticism actually points to an important subtlety. If a signature is present for a change of regime in the underlying dynamics of a market (or a material), the proper conclusion is that the risk profile of sustainability has changed, not that a rupture is a certainty. When such a signature appears – as in changes preceding earthquakes – the prudent course of action, considering the potential (probability-adjusted) cost of inaction, is to evacuate even knowing that the tremor may subside. A 'financial bubble experiment' is underway to rigorously test the predictive power and the limits of the proposed methodology (Sornette et al., 2009 and www.er.ethz.ch/fco).

The motivation of the Financial Bubble Experiment (FBE) comes from the failure of standard approaches to financial crises. Neither the academic nor professional literature provides a clear consensus for an operational definition of financial bubbles, nor for techniques to diagnose them in real time. Instead, the literature reflects a permeating culture that simply assumes that any forecast of a bubble's demise is inherently impossible.

The FBE aims at testing the following two hypotheses:

- *Hypothesis H1*: Financial (and other) bubbles can be diagnosed in real-time before they end.
- *Hypothesis H2*: The termination of financial (and other) bubbles can be bracketed using probabilistic forecasts, with a reliability better than chance.

Because back-testing is subjected to a number of biases, the FBE is proposed as a real-time advanced forecast methodology that as far as possible is free of the biases plaguing previous tests on bubbles. In particular, active researchers are constantly tweaking their procedures, so that predicted 'events' become moving targets. Only advance forecasts can be free of data-snooping and other statistical biases of *ex post facto* tests. The FBE aims at rigorously testing bubble predictability using methods developed in our group and by other scholars over the last decade. See Sornette et al. (2009) for more information on the specific procedure. Two conclusions can be drawn from such a model: that a bubble in fact exists, and that there is probabilistic window for its end.

Policy-Making in the Aftermath

For many years, the consistent philosophy of United States Federal Reserve Bank officials and their counterparts in other market-based economies was to do nothing to bubbles developing in the pricing of stocks, houses or other tradable assets (currency proper and foreign exchange largely excluded). This approach was rooted not so much in unadorned free-market economic theory as in a mistaken trust in the effect of prior regulatory structures. It also points to a subtler difficulty in dealing with bubbles. It is widely believed to be in the nature of a bubble that it is generally not identified as such until after it ends, which means that its abrupt and damaging burst confirms its existence.

The best explanation for this widely discussed phenomenon is perhaps the simplest. A minority of individuals might correctly guess that a continuing run-up in asset prices is being driven primarily by the prior history of the run-up itself (in contrast to a hypothetical increase in 'fundamental' value). An even smaller minority of these might be willing to act on their guess either by withdrawing or shorting the market. But by definition a market-weighted majority of investors must act as if they did not believe a bubble was inflating, or at least that prices were not so 'irrational' vis-à-vis 'fundamental valuation' that a collapse was imminent. This explanation also reflects a substantial degree of efficiency in a market.

One might argue that bubbles by definition are a phenomenon characterised by being mischaracterised. A widely recognised bubble cannot develop precisely because it is accepted as such. There is no simple, self-consistent exit out of this potential infinite regress. Note that this argument tacitly presumes that a common underlying mechanism drives prices (hence draw-downs and draw-ups) at all scales. If this is not true, if regime changes do occur and carry a signature, then detection of a bubble may be possible. As Ben Bernanke testified in his December 2009 reappointment hearings as Chief of the US Federal Reserve, financial booms are 'perhaps the most difficult problem for monetary policy this decade'. This statement is actually an understatement, since financial booms have been and remain the most difficult and important problem of the last 100 years and more (Reinhart and Rogoff, 2009). 'The best approach here, if at all possible, is to use supervisory and regulatory methods to restrain undue risk-taking and to make sure the system is resilient in case an asset price bubble bursts in the future', Mr Bernanke said in a November 2009 speech. The Bank for International Settlements in Basel, Switzerland, coordinates central-bank activities around the world, and is likewise pushing to have bubbles confronted more aggressively.

William Dudley, former chief economist at Goldman Sachs and now president of the New York Federal Reserve, is one of the more outspoken proponents of preventing bubbles, and has said that it is not as hard to spot them as many economists believe. 'I can identify at least five bubbles that one could reasonably have identified in real time', including the tech boom, Mr Dudley said in his speech in April 2010. He knew, he said, because he had 'speculated against three of them himself when he was at Goldman' (Dudley, 2010:1–8). This resonates with the analysis of five financial bubbles in the beginning of this chapter (Sornette and Woodard, 2010).

It is important to understand the close relation between underappreciated dependencies of the kind that led to the MBS debacle and the underappreciated dependence over time in asset price bubbles as they enter the high-risk regime. Tobias Adrian, a New York Federal Reserve researcher, and Hyun Shin at Princeton show that the credit bust was preceded by a sudden burst of short-term borrowing by Lehman Brothers, Bear Stearns and others (Adrian and Shin, 2007). Borrowing in the 'repo' market, where securities are used as collateral against short-term loans. This short-term borrowing more than tripled, from US$0.5 trillion in 2002 to US$1.6 trillion in 2008. The 'value' of the collateral rose, allowing increased borrowing against it. This is an example of the positive feedback discussed above, but in this case between two related investment classes.

Leverage, its interconnectivity and positive feedback loops are major seeds for destabilising a Financial Services system. It is possible that a very dangerous part of an asset price bubble consist not only in the inflated prices, but in the level of debt that helps fuel the rise. This scenario, together with our earlier analysis of correlation in MBS security tranches, suggests that one kind of intervention that might be effective when the signature of a bubble appears is to cap or even reduce allowable debt levels, that is, control leverage. This conclusion has been reached by several other economists, who argue that the problem of the financial crisis was not a liquidity crisis (which is only the short-term consequence) but a fundamental problem associated with over-leveraging.

Ultimately, the focus on the liquidity crisis and credit freeze is misleading; the fundamental problem is leverage and its unanticipated impacts on interconnectivity, interdependence and positive feedbacks, which plant the seeds of systemic instabilities.

The use of leverage implies a high level of confidence that a financial market will not undergo an abrupt phase transition and suddenly begin to behave differently, with a different risk profile. The design of complex *synthetic* financial derivatives implies a high level of certainty that the underlying real assets are unlikely to share correlated risk. The present understanding of how assets are priced by markets also assumes that the efficient market hypothesis is essentially correct, and that positive feedback loops do not occur within price time series. All three assumptions are very often false and greatly underappreciated. The result is a very poor understanding of risk.

Spontaneous Recovery

Another factor to be considered is that while extreme risk and actual crashes usually are an unanticipated feature of financial markets, just as attempts to avoid them usually are subject to an illusion of control, the converse situation is equally a feature of financial markets. That is, out of the apparently uniform rubble post-crash, markets show a surprising spontaneous resilience. In an important paper presented to the National Academy of Sciences of the US in 2000, Shnerb et al. (2000) demonstrated that the typical continuum, representative agent or mean-field approach to analysing financial markets (and other complex systems), often predicts that invested capital entirely vanishes under adverse circumstances.

In this continuum, representative agent or mean-field approach, the collective dynamics of all individual components of a system are approximated by equations operating on continuous spatial density distributions. This approach makes the intuitively plausible assumption that the statistical approximation by a smooth field (or even a single 'representative agent') does not introduce a significant distortion for a large enough number of spatially distributed interacting agents. In fact, however, local granularity allows for significant sub-populations that resist the general decline, and nucleate emergent success and a new round of growth and development. Successful use of resilience engineering principles may thus require both a greater appreciation of seemingly unimportant local variations in a system and greater care when using mean-field approximations. In an extension of this theoretical perspective, and reflecting our earlier comments about Dragon Kings, only this time positive ones (the equivalent of 'draw-ups' instead of 'draw-downs'), Yaari et al. (2008) demonstrated that the dramatic growth in Poland during the 1990s was stimulated and dominated primarily by rare, localised 'growth centres' that 'initially developed at a tremendous rate, followed by a diffusion process to the rest of the country'. An analysis that averaged over the country as a whole 'blurs the singular character of the growth centers which is the dominant feature'. Yaari et al. (2008) have thus used the post-liberalisation empirical economic recovery of Poland to offer a remarkable empirical validation of the singular growth autocatalytic model.

In addition to intervention aimed at predicting and reducing excessive risk, we suggest that in particular post-crash governments should pay close attention to emerging centres of growth and development and support them. It is perhaps counter-intuitive to be seen rewarding those already demonstrating evidence of success, but the above finding suggest that overly aggressive attempts to equalise wealth, or wealth production, could have the effect of creating a dangerous 'mean field', not as an approximation but in reality. The perverse result of such well-intended actions could well be to suppress the wealth-generating mechanism that has the best chance of raising the well-being of the entire context.

Chapter 7

Using Power Laws and the Hurst Exponent to Identify Stock Market Trading Bubbles

Rossitsa Yalamova and Bill McKelvey

The efficient market hypothesis (EMH; Fama, 1970) implies that normally functioning financial markets, that is, markets in a normal state, are able to efficiently incorporate random information into stock prices. Markets may over-react or under-react to specific bits of information, but price deviations are presumed short-lived and normally distributed. A dynamic equilibrium between demand and supply keeps prices fair and markets free of arbitrage. In this market state, traders act independently and their actions are heterogeneous.

These efficient market dynamics are a result of the collective behaviour of independent traders who make rational decisions without being influenced by the decisions and actions of other traders. Traders manage various amounts of money and obviously have different investment time horizons. The efficient market hypothesis holds that a dynamic equilibrium is achieved by market prices varying randomly around the 'Triple Point' basin of attraction. This so-called Triple Point is defined by information, risk, and uncertainty and by short-lived anomalies (Yalamova and McKelvey, 2011).

Yalamova and McKelvey (2011) propose a phase-transition model of market dynamics that allows for extreme events and explains the origin of power laws in stock price volatilities as well as power laws of the autocorrelation function. Autocorrelation is a mathematical tool that enables measurement of the similarity between observations as a function of the time separation between events.

In essence, the proposed model accounts for the increased information-complexity conditions (and/or noise) that impedes rational decision-making. In these kinds of situations traders resort to imitation and herding and tend to follow similar rule-based trading approaches (that is, traders buy/sell based on converging logic and/or beliefs rather than independent assessments of fundamentals). This financial market phenomenon is well documented empirically (Banerjee, 1992; Bikhchandani et al., 1992; Brunnermeier, 2001; Hirshleifer and Teoh, 2003; Rook, 2006).

Above the noise threshold, market self-organisation facilitates cluster-formation of traders, whose activities are synchronised in such a way that a dominant trading rule might emerge and as a result lead to a crash as illustrated by LeBaron (2001). Our proposed model also explains the pre-crash log-periodic patterns of index prices described by different groups of physicists (Yalamova, 2003; additional references can be found in Introduction to Section II).

The Yalamova/McKelvey phase-transition model calls for different research methods in the two distinct market states: (1) the efficient-market state assumes random changes in stock prices and normal distributions of returns, as well as fast decay of the autocorrelation function; and (2) the market state beyond the previously discussed information complexity threshold, characterised by behaviour resulting in price-volatility power laws and a slow decay (long memory) of the autocorrelation function.

Financial Markets – Toward a Scale-Free Perspective

Sornette et al. (1996) compare the three phases of matter (solid, liquid and gas) to the three possible trader states of wait, buy and sell. Our model uses a phase diagram and defines not the individual trader's state but the aggregate state of the market as 'wait', 'buy', and 'sell' as a function of the balance of the order book (net order balance). Given a state of the market where demand equals supply, 'hold or wait' is characterised by small price changes. More demand than supply corresponds to positive net demand; we call this state 'buy', while a negative demand state characterised by more supply than demand is called 'sell' (see Figure 7.1).

We begin with short definitions of market dynamics as defined by EMH (Fama, 1965, 1970, 1998), which we label as the *Triple Point* in the phase diagram. Then we define the *Critical Point*, where market crashes occur, and finally, we focus on how markets build up from a Triple to Critical Point.

Triple Point

The Y-axis in the market phase diagram (see Figure 7.1) measures the risk of a security as incorporated in the fundamental valuation. Closer to the Origin, the risk measure approaches zero, leading to overly cautious evaluation of the fundamental values of firms, which leads to underpricing of stocks. At the point of equilibrium, that is, at *TPy1*, risk is properly incorporated in the fundamental analysis and the price is 'fair'. As risk increases, risk-taking

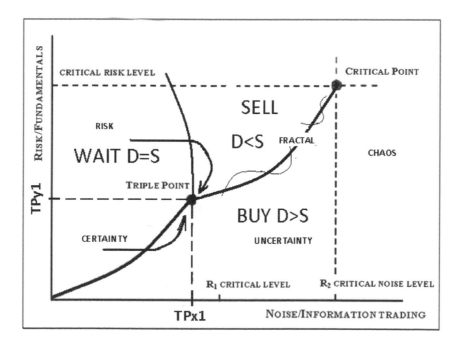

Figure 7.1   Financial markets phase diagram

Source: Reproduced from Yalamova and McKelvey, 2010.

behaviour becomes more dominant and underestimation of the volatility of a firm's underlying fundamentals becomes more frequent, resulting in overpricing.

Rational investors evaluate all available information and make their trading decisions based on the ratio of Risk and Return, defining underpriced (buy) and overpriced (sell) stocks. The Y-axis of our financial markets phase diagram captures these decisions by a Risk/Fundamentals ratio. Sell pressure on the price increases with increase of Risk (on the Y-axis), that is, investors are predominantly risk averse. In normally functioning markets, that is, where system complexity is minimal and rational traders hold uniform assumptions, the opposite side of a trade is attributed to liquidity traders in the efficient-market paradigm. We also allow for noise traders either with 'incorrect' valuation, or invalid information, to make decisions to sell underpriced and buy overpriced securities. As we will explain later, the increasing number of such noise traders will lead to a bubble build-up.

On the X-axis, we focus on the Noise to Information trading ratio. We believe that disorder in the market can be measured by the ratio of Noise to Information trading. Information trading implies that investors can properly process information and act rationally. Close to the origin, information-based trading and rationality prevail. If all investors are rational and all trading decisions are information based, we will have homogeneous agents, one-sided trade orders, and a 'halt' in the market. At *TPx1* a balance between information and noise traders sets the market at fair valuation. Conversely, as noise trading becomes increasingly evident in the market, buying and selling behaviour based on information and proper valuation procedures are replaced by imitation, information cascades and herding behaviour.

In normal markets, with heterogeneous independent traders and random noise, prices adjust to fundamental or certain values; anomalies are short-lived. Under some conditions (for example, new technology, trading rules, or formulas such as derivatives, and so on), noise in the market increases and creates information ambiguity. Therefore uncertainty-averse traders switch to non-probabilistic decision-making approaches.

Moving horizontally from the area of 'certainty' to the right, we enter a region of higher information complexity and uncertainty 'distinct from the familiar notion of risk', as defined by Knight (1921). In uncertain situations many decision-makers prefer to bet on unambiguous events rather than on ambiguous ones (Basili and Zappia, 2003). Shackle (1949) developed a theory opposing the subjective probability approach. It is a non-probabilistic decision theory seeking to optimise robustness to failure, or opportunities for windfall profits. Zhang (2006) investigated the role of information uncertainty in price continuation anomalies and cross-sectional variations in stock returns; he showed that short-term price continuation was due to investor behavioural biases, which resulted in greater price drift when there was greater information uncertainty. We argue that, given the random nature of good/bad news, information uncertainty alone cannot produce a bubble-build-up, since greater information uncertainty should produce relatively higher expected returns following good news and relatively lower expected returns following bad news.

Triple Point Dynamics

The lower-left corner of the phase diagram in Figure 7.1 is an area of system instability; that is, rational investors are willing to buy the underpriced securities; however, there are very few noise traders, willing to sell at this price. As a result, the market moves from left to right, thereby crossing the phase boundary (the heavy line in Figure 7.1); this results in falling liquidity as buy orders exceed in the next phase and prices increase. The basin of attraction is the *Triple Point*. Note that in the 'certainty' area, information traders dominate the market.

The upper-left corner of the phase diagram is an area of overpricing based on higher risk taking. Note, however, that news about fundamentals may cause higher volatility leading to correction. Rational investors with adequate valuation will detect overpricing and place sell orders when the market rises above *TPy1*. Since the only traders who buy under these conditions are noise traders, the market moves to the right, and crosses the phase boundary into the 'sell' phase; liquidity falls, uncertainty increases, and prices adjust back down to the equilibrium market risk levels at both *TPx1* and *TPy1* – which is our Triple Point.

Over- or under-pricing is a very short-lived phenomenon after new relevant information is released. In both cases, that is, bad or good news, moving from the 'wait' phase through sell or buy to reach the attractor point (that is, the Triple Point, which represents dynamic equilibrium), the market crosses the phase boundaries. If nonlinearity is present, there should be a function that experiences abrupt change with a small change in the X-axis (noise/info trading). This also should be related to a jump in the price, as the market quickly adjusts to new information and incorporates the news in the price.

As we move horizontally in Figure 7.1, from 'wait' to 'buy' in the area of under-pricing, liquidity increases due to the presence of more noise traders. When we cross into the 'buy' state characterised by positive demand, the liquidity function changes. Similarly, in the overpriced area, when a rational investor wants to sell at *x* price, the market should move horizontally in search of 'noise' buyers. To the right, the ratio of noise traders' increases, ease of trade increases, that is, liquidity increases. After crossing the phase boundary, the market enters the area of negative demand and as a result sell orders have fewer buyers, that is, liquidity decreases. According to the EMH, these are short-lived anomalies that are arbitraged away. The simultaneous execution of a large number of trades produces efficient outcomes and a dynamic equilibrium between the three states is present. Also according to the EMH, all investors are rational and base their decisions on fundamental values; the opposite side of the trade is taken by liquidity-need investors.

From Triple to Critical Point

Having shown the Triple Point to be an effective attractor basin, we then need to find a mechanism that explains the reality of extreme price volatilities, that is, bubble-build-ups and crashes. Bubble-build-ups and crashes, producing extreme returns, appear much more often than the Gaussian distribution predicts (Sornette, 2003b; Mandelbrot and Hudson, 2004; Baum and McKelvey, 2006). A normal market efficiently incorporates new information into the stock prices, and mispricing of securities is temporary. Information does not create ambiguity, noise in the market is offset by rational decision-making, and anomalies are short-lived. When noise levels increase, information ambiguity prevails and market moves to the right in the region of uncertainty, efficient information processing and probabilistic decision-making become more difficult.

Grossman and Stiglitz (1980) argue that if information gathering is costly, a competitive Walrasian market does not always remain in equilibrium. Moreover, Grossman and Stiglitz also illustrate that even when the EMH is correct, costly information can cause competitive markets to break down. Informed traders realise that they can stop paying for information and still do as well as uninformed traders. Therefore, having some fraction of informed traders does not necessarily produce equilibrium. Having no one informed does not produce an equilibrium either, if each trader thinks that there are profits to be made from becoming informed.

Following Grossman and Stiglitz, we do not reject the EMH but rather wish to extend it to fit conditions when information is complex, that is, costly and/or problematic market thinking with respect to market analysis. If information complexity increases, analysis is more costly and imitative trading (herding) is more attractive, if not temporarily optimal. When the level of noise increases, information ambiguity prevails and the market is pervaded by uncertainty; efficient information processing and probabilistic decision-making becomes more difficult.

The region of noise-dominating uncertainty, characterised by the complexity and ambiguity of information, begins at $R1$ and extends to the right – $R1$ marks the transition from efficient market trading to herding behaviour. Rational probabilistic decision-making is impeded and the bimodal-demand function emerges in the 'limit order book' data. At $R1$, information complexity impacts risk-taking behaviour such that a market moves to a higher risk/fundamentals area. Information cascading, herding, rule-based trading and so forth create a complex self-organised network among traders and leads to a power-law distribution of price volatilities (crashes being examples of extreme volatility or price movements). Therefore we define this area as 'fractal'. It is characterised by Pareto distributions and power-law distributions when the data are plotted on double-log scales (for further definition see Newman, 2005; Andriani and McKelvey, 2007; see also Chapter 4).

Sornette (2003a,b) presents a general theory of financial crashes and stock market instabilities and asserts that markets exhibit complex dynamics. Moreover, he suggests that large-scale patterns of a catastrophic nature result from global correlated trading processes caused by repetitive interactions that eventually spread across the entire system. A power-law distribution punctuated with log-periodic oscillations in the index prices seems to be the signature of an impending crash. Among many other examples, Baum and McKelvey (2006) also show evidence of power-law distribution in the daily log returns of Dow Jones and NASDAQ. They argue that observed power laws stem from interconnected behaviours that are ever present in social contexts and consequently also present in stock markets.

The Four Regions

Having defined the Triple and Critical Points, we now have four regions in Figure 7.1: Certainty, Uncertainty, Risk, and Fractal. Each region is defined as follows:

1. *Certainty*. Most pronounced at the *Origin*. Note the points on each axis where the Risk/Fundamentals and Noise/Information ratios equal 1, that is, *TPy1* and *TPx1*. On the X-axis, accurate information dominates noise. On the Y-axis, risk measured by the volatility of a firm's underlying fundamentals is low relative to that used in the valuation process. Rational traders unanimously agree on the fundamentals-based pricing the closer to the Origin the market moves – subject to the random movements characterised by the EMH.

2. *Uncertainty*. To the right of the Triple Point location on the X-axis, traders lose any reliable means of attaching true value to information about a particular stock/company – noise therefore dominates. Complexity and ambiguity of information take over the market. While uncertainty keeps increasing towards *R2*, the market becomes vulnerable to chaos – that is, bifurcations due mainly to random external anomalies hitting a market. Price continuation anomalies are increasing as uncertainty levels increase.

3. *Risk*. Above the location of the Triple Point on the Y-axis, traders move away from only trading based on knowledge about the *current* 'fundamental' value of a stock/firm to start betting on *future* value. A rational bubble emerges when market price depends on its own expected rate of change reminiscent of the models of Blanchard (1979) and Blanchard and Watson (1982). Risk increases up to the location of the Critical Point on the y-axis. Above this point we show 'Chaotic Risk'; this is the point where risk-taking becomes vulnerable to chaos – random bifurcations that can set off significant crashes.

Note that we show Knight's (1921) risk, uncertainty, and certainty as juxtaposed at the Triple Point. This is the core explanation underlying the EMH – traders leaning toward all three situations trade concurrently with quick adjustments of the market shifting toward one or the other of the three conditions.

4. *Fractal*. The Region between the Triple and Critical Points is notable for increasingly dramatic price-volatility incidents. Since there is growing evidence that many of these incidents follow fractal patterns, we label the region Fractal. This region corresponds to the bubble-build-up regime (Sornette 2003a; Yan, Woodard and Sornette, 2010) where non-stationary increasing volatility correlations are reported. Moreover, regime switches between 'normal' and 'bubble' comprise a dynamical model that recovers all the stylised facts of empirical prices. This is what we focus on next.

How much time does the Dow Jones, for example, spend in the fractal region? Note from Figure 7.1 that, as risk and uncertainty increase, traders end up in the fractal region. 'Tradition' and the EMH in Finance hold that it spends most of the time at the Critical Point. For empiricists this is represented by GARCH (Generalised Autoregressive Conditional Heteroskedasticity; Bollerslev, 1986). But as one can see in Figure 7.2, there are many market variances well above the 'GARCH line' indicated by the slightly darker line in Figure 7.2.

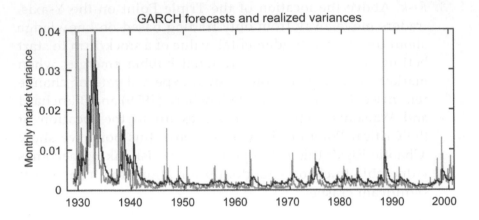

Figure 7.2   Depiction of volatility incidents above GARCH line

Source: Reproduced from Ghysels et al., 2005.

Mandelbrot calculates that:

> by the conventional wisdom, August 1998 simply should never have happened
> ... The standard theories ... would estimate the odds of that final, August 31,
> collapse, at one in 20 million – an event that, if you traded daily for nearly 100,000
> years, you would not expect to see even once. The odds of getting three such
> declines in the same month were even more minute: about one in 500 billion (p. 4)
> ... [An] index swing of more than 7% should come once every 300,000 years; in
> fact, the twentieth century saw forty-eight such days. (Mandelbrot and Hudson,
> 2004: 13)

In fact, market behaviour *around* the Triple Point and *between* Triple and Critical Points exists. For sure, on a daily basis efficient market behaviour is more abundant. But market behaviour in the Fractal region is also present and often with greater consequences. The number of days the market spends at the Triple Point is considerable. Though the number of days in the fractal region is far fewer, the costs stemming from what happens in this region are often very high indeed – the financial losses and unemployment during the current Great Recession are dramatic illustrations of this.

Fractals and Complexity

Fractals, power laws, and scale-free theory were defined in Chapter 5, which also included a table connecting various scale-free theories to various historical inventions and events occurring in financial transactions and stock trading since 1972. In this Section we introduce some of the evidence showing that herding behaviour exists during the bubble-build-up region of Figure 7.1 – the region between the Triple and Critical Points.

Power-Law Distributions of Returns

When the EMH applies, prices follow random walks and become unpredictable as their changes efficiently incorporate randomly arriving information. Stock-price changes measure returns that follow a Gaussian distribution. A stock 'return' is the money gained or lost when a stock is sold. Since 'return volatilities' are based on 'price volatilities' we mostly refer to price volatilities in the present chapter.

There is sufficient empirical evidence to illustrate that extreme events (that is, those beyond three standard deviations), are frequent and as a result should not be categorised as random anomalies (Mandelbrot and Hudson, 2004). Mandelbrot and Hudson point out that daily stock trading returns of seven standard deviations from the mean, occur 10,000,000 times more frequently than would be the case for random data. Our proposed model allows for efficient markets with Gaussian returns at equilibrium and self-organisation and power-law distribution of returns in a regime with increased information complexity.

The presence of power-laws in price volatilities points to interesting dynamics. We argue that 'dynamics' are not appropriately represented in modelling financial data if emphasis is primarily on standard statistical research methods. Needless to say, for our purposes, information about the distribution of past events is not very useful from a resilience engineering perspective. We suggest that power laws provide more useful information with a more solid empirical basis.

The distributions of a wide variety of events seem to follow the power-law form (that is, the log of the frequency of events decreases while the log of size increases, cf. Chapter 5). The behaviour of the Pareto extremes lying behind power laws is the rationale of theories explaining extreme deviations (for example, extreme-value theory; Baum and McKelvey, 2006), which are

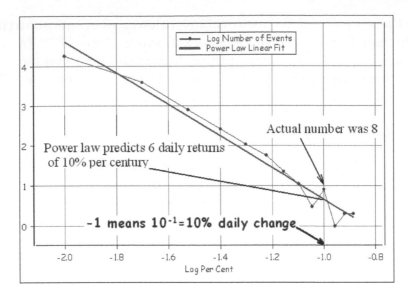

Figure 7.3   Power-law fit DJIA (# events vs. per cent daily change)
Source: Reproduced from Keen; no date.

focused on identifying the probability of extremely rare events like stock market crashes and large natural disasters. The most popular way of showing power-law distributions of returns is to plot the log of the number of observations against the log of the return size and thereby produce an inverse sloping line that can be used to indicate a power law. Figure 7.3 illustrates this method by showing a power-law distribution of the stock-price changes in DJIA.

Although the field of Finance is built on the premise of the time-value of money and inter-temporal choices, our research reveals that little attention is given to the temporal conjunction of volatility events. Our proposed model requires an approach that further validates the emergence of complex structures and dynamics leading to the *Critical Point* and financial market crashes.

Power Laws of the Autocorrelation Function (ACF)

Financial time series consist of observations of a stock-price change process that may be non-stationary, that is, the probability distribution changes as points in time advance. Thus, in order to transform the data into stationary series, one needs to calculate the degree of integration, which determines how many differencing steps it will take to transform the data. For example, a first-order integrated process becomes stationary after taking the difference between each observation and the previous one, that is, $X(t) - X(t–1) = 1$. After this step is taken, the series become stationary and can therefore be analysed using tools designed for random processes. If the differencing step is an integer (1, 2, 3, and so on) it appears to create no contradiction within the realm of efficient market (E-M) trading behaviour. The prices follow a 1st, 2nd, and so on, order integration process for which econometric tools are widely accepted.

The standard assumption of an integer-integration order is arbitrary. Allowing for fractional-integration order leads into the world of fractals that creates the division between the statistically 'normal' trading behaviours assumed by the EMH and the 'complexity' of fractal models. In a fractal market regime, the value of returns is still not predictable; however, volatility *is* predictable (Poon and Granger, 2003). Classic econometrics shows the clustering of volatility; however, it does not capture long memory in the volatility of returns unless it allows for fractional integration in the models, that is, the differencing step can be fractional as opposed to only accepted as integer.

If we have a non-stationary process that becomes stationary after differencing, and the differencing operator is fractional, we enter the realm of fractals. This is a sign that structural changes are taking place. In a financial market this will appear in the slow decay of the autocorrelation function (that is, long memory in volatility). The hyperbolic (slow) decay of the autocorrelation function indicates dependence, correlation, and feedback between traders; that is, imitation, herding, and rule-based trading. Our hypothesis is that when information becomes expensive and/or complex and ambiguous, the market cannot quickly transition back to the Triple Point, that is, back to the equilibrium state. The long memory in stock-returns volatility has already found its way into econometrics models (Granger and Joyeux, 1980). Given that many time series exhibit very slowly decaying autocorrelations, the advantages of Autoregressive Fractionally Integrated Moving Average models with hyperbolic autocorrelation decay is clear.

The stationary stochastic processes frequently referred in financial time series, such as ARCH (Engle, 1982), GARCH (Bollerslev, 1986), IGARCH (Engle and Bollerslev, 1986), and EGARCH (Nelson, 1991) all presume short memory for volatilities.

Long memory is defined mathematically in terms of autocorrelation. The time series of the absolute value of returns exhibits autocorrelation, when return $|r_i|$ is correlated with $|r_{i-s}|$ and s is a measure of time lag.

If time series exhibit long memory, they display significant correlations between observations separated in time – that is, the correlation does not go to zero given a long lag. Therefore we focus our discussion around the evidence of increased price volatility persistence in periods before crashes. Translated into the language of power laws associated with extreme events, we point to the appearance of power laws in the autocorrelation function (ACF) of volatility, as illustrated in Figure 7.4.

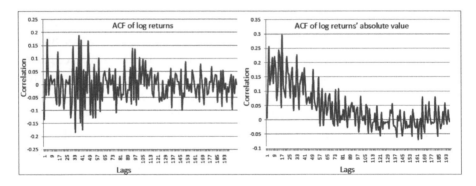

Figure 7.4   DJIA daily returns 1928–2007

Source: Reproduced from Yalamova et al., 2007.

The ACF of log returns is effectively zero for lag > 2, while the ACF of log-return absolute values, as a straightforward measure of volatility, decreases very slowly and remains positive, even after a lag of 500, before decaying to zero. Such results suggest that though the returns of a stock market are uncorrelated, their volatility has long-range dependence. A logarithmic-scale plot of the autocorrelation function gives the same visual illustration of power laws as described in our explanation of power-law distributions above.

The autocorrelation function ϱ takes a power-law form with constant c and exponent α:

$$\rho(s) = cs^{-\alpha}$$

The fractional integration parameter α of the ACF is related to the generalised Hurst exponent (H) as:

$$H = 1 - \alpha/2$$

H is a measure of the extent of long-range dependence in time series. H takes on values from 0.5 to 1. A value of 0.5 indicates the absence of long-range dependence. The closer H is to 1, the greater the degree of volatility persistence.

Long-range dependency relates to the rate of decay of statistical dependence measured by the autocorrelation function, with the implication that slower than an exponential decay typically means a power-law indicated decay rate.

Volatility is random if the Hurst exponent is equal to 0.5, which indicates totally random, so-called 'Brownian motion' movements – which characterise E-M trading behaviour. Volatility persistence as measured by the Hurst exponent increases above 0.5 during periods of increased information complexity. More specifically, if $H > 0.5$, rule-based trading, herding, imitation, and increased mutual influence among traders leads to log-periodic oscillations of prices appearing as precursory bubble-build-up patterns before crashes, that is, during the build-up between Triple and Critical Points as illustrated in Figure 7.1.

The self-organisation process causing the foregoing market dynamics appears in the form of power-law distributions of returns as well as in a power-law of price volatilities. Yalamova (2003) illustrates the beginnings of increasing persistence in volatilities as measured by the Hurst exponent in a number of stock market indices in periods of 2–4 years *before* significant draw-downs (crashes) such as those that occurred in October 1987 and March 2000.

Identifying Fractality via the Hurst Exponent

Hierarchical structures appear as fractals when, as in a cauliflower, the shape and function of the whole appears essentially the same in smaller subunits, right down to subunits we can barely see. Although the mathematics of fractals dates back to the nineteenth century, the term was introduced by Mandelbrot in 1975. More recent treatments appear in Mandelbrot (1982, 1997); Schroeder (1991); Peitgen, Jürgens and Saupe (1992), Peters (1994), West and Deering (1995), and Sornette (2003b, 2004).

The Euclidean dimensions appear as 0 for point, 1 for line, 2 for plane, and so forth, whereas fractal dimensions allow us to measure the degree of complexity by evaluating how fast our measurements increase or decrease as we go from a whole down to its smallest elements or parts. Hence, a defining property of a fractal is 'self-similarity', that is, the whole has the same appearance and underlying causality as one or more of its parts. A straight-line power law is a reliable method for identifying well-formed fractal structures and Pareto distributions, which we find growing evidence of in nature and in social and financial phenomena (Newman, 2005; Andriani and McKelvey, 2007, 2009; Chapter 5, this volume).

The Hurst exponent was originally developed in hydrology for the practical matter of determining optimal dam-sizing for the Nile River's volatile rain and drought conditions, which had been observed over a long period of time. The Hurst exponent is non-deterministic in that it expresses what is actually observed in nature; it is not calculated so much as it is estimated. Thus:

$$H = \frac{(slope - 1)}{2}$$

The Hurst exponent is related to the fractal dimension, which gives a measure of the self-similarity of a surface (Struzik, 2001; Yalamova, 2003, 2010; Grech and Mazur, 2004; Cajueiro and Tabak, 2004; di Matteo et al., 2005; Maskawa, 2007; Alvarez-Ramirez et al., 2008; Eom et al., 2008). It has emerged as a reliable indicator of the $R1$ tipping point between randomly occurring anomalies in efficient market and the power law defined bubble-build-up leading to the Critical Point and a market crash. The relationship between the fractal dimension, D, and the Hurst exponent, H, is:

$$D = 2 - H$$

Information and Noise

Rational traders in an efficiently functioning market process freely available information and make trading decisions based on fundamental values as described in the context of our *Triple Point* in Section 1. We also describe the dynamics fostering bubble-build-ups to the *Critical Point* that may be set off in situations of costly information (for example, introduction of new technology, complex derivative instruments, and so on) leading to herding behaviour that, at least temporarily, obscures the increasing fractions of noise- and risk-based trading, which head the market towards a new attractor point, and perhaps even to the Critical Point.

'Tipping Point' R1 – Information Complexity Threshold

The high variability (volatility) of stock market prices is a signature of collective phenomena such as imitation or 'herding' behaviour (Banerjee, 1992; Bikhchandani, et al., 1992; Brunnermeier, 2001; Hirshleifer and Teoh, 2003; Rook, 2006). A quantitative link between bursts of volatility and herd behaviour can be established through examination of order flows that display the aggregation of individual demands independent of the mechanism of herding, whether it is a sequential information cascade or random formation of groups through clustered networks. Implicit in herding behaviour is the slow consolidation of traders toward the same buy–sell rule, that is, rule-based trading, which then leads to correlated behaviour.

In an environment of increased uncertainty, rational probabilistic decision-making is impeded and the bimodal-demand function (Plerou et al., 2003) signals herding behaviour. At a certain level of information complexity, or when the buy/sell rule of another trader or group of traders is leading to obvious financial gain, a trader usually resorts to the alternative of rule-based trading instead of behaviour consistent with the EMH. Rule-based trading, herding, and information cascading, and so on, create complex networking (self-organisation) among traders.

Under such circumstances, noise trading that disrupts normally functioning markets has increased impact because noise can disrupt the symmetry among demand/supply and destabilising prices, that is, disrupt Triple Point trading. Below the critical value of information complexity, the net demand is roughly zero; neither buying nor selling predominates, which agrees with the dynamic stability in the basin of attraction, that is, our Triple Point of a normal efficient market at equilibrium.

Above this critical-noise tipping point, $R1$, a bimodal distribution of buy and sell orders emerges with the two most probable values symmetrical around zero demand as reported by Plerou et al. (2003). This empirical evidence of change in the net-demand distribution explains the frequent reversal of the market, which creates the oscillation of prices, as documented by Sornette (2003a,b). Sethi (1996) shows that if chartists control a significant portion of trading, destabilised prices may show a sequence of period-doubling bifurcations. In this regime, markets – as complex dynamical systems – exhibit oscillations showing log-periodic behaviour as discovered empirically by Sornette et al. (1996), and Johansen and Sornette (1998), among others. The bimodal distribution of demand reported by Plerou et al. (2003) suggests oscillation of the market between negative and positive demand phases in our phase-transition diagram (Figure 7.1). A phase transition at this, the Triple Point, is related to abrupt changes in the trading volume, induced by liquidity constraints. The reversal frequency of market sentiment is related to the increasing hazard rate of crash-producing log periodicity in price oscillations. In the bubble-build-up sequence, rational traders evaluate the increasing price trend and hazard rate and then adapt their speculative strategy accordingly. Sornette (2003a) describes the *build-up of cooperative speculation*, which usually translates into an accelerating rise of the market and price volatility.

Grech and Pamula (2008) show that shortly before a crash a few of what they call 'speculative' traders start abandoning the herd; as more begin to jump from the bubble, the period of log periodicity ends and a crash occurs. But, at some point there is a likelihood that what starts as an information cascade becomes, itself, more resilient such that only significant outside anomalies can disrupt it. For example, in a recent study of the S&P and NASDAQ stock markets at the times of the 2000 dot.com bust and the 2007 liquidity crisis, Fagura, Kunal and Poag (2010) find that the crashes comes 1 to 1.5 years after the R1 tipping point. *This suggests that there is a fair amount of time between R1 and crash for the resilience interventions to be imposed such that they counteract the more dramatic market crashes, cf. Chapter 9.*

The foregoing may be illustrated as follows: in Figure 7.5 we show Sornette and Johansen's (2001) translation of the Hang-Seng stock chart from linear to log Y-axis. The black line in the graph shows the log-scaled upward trend – the market rises exponentially at an 'average' rate of 13.6 per cent. Each period between a crash and following peak has been bifurcated by the intersection of two lines: (1) the *white* lines, which indicates the period of trading at the Triple Point via E-M (Hurst exponent ≈ 0.5); and (2) the steeper-angled *dashed white* lines, which indicates the period of log-periodicity, persistence, and (Hurst exponent > 0.5).

When information complexity in a market breaks across the R1 tipping point (bifurcation point), deviations from the Triple Point 'fair' prices may not be readily and efficiently corrected, given the presence of large numbers of rule-based traders, which may then set off a bubble-build-up. Participants in the market influence

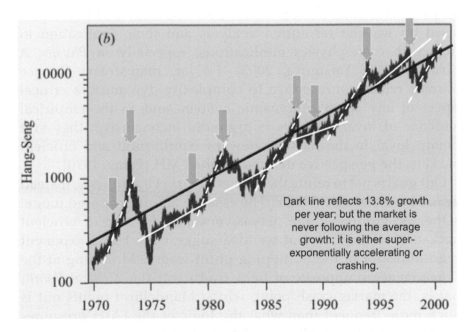

Dark line reflects 13.8% growth per year; but the market is never following the average growth; it is either super-exponentially accelerating or crashing.

Figure 7.5   **Log-periodicity in the Hang-Seng stock market (1970–2000)**

each other in their trading decisions. Independent rational decision-making gives way to noise, copying, and imitation; the independence and heterogeneity of agents that keeps the market in equilibrium is lost. Traders self-organise in larger and larger clusters; finally the traders' network synchronises resulting in a crash. Sornette (2003a,b) describes this as order in the market when everybody has the same opinion, 'sell'. LeBaron (2001) finds that traders losing their diversity of buy/sell rules evolve to one dominant trading rule, a finding that also supports our concern about finding the tipping point that sets herding behaviour in motion, which can lead up to a crash.

In summary, Classical Finance Theory and asset pricing models are mostly based on traditional statistical research methods. Econophysics attempts to adopt a more complex approach to modelling that considers not only statistics but also dynamics and temporal allocation of events. Empirical research in Finance will benefit tremendously from the application of methods that incorporate time and scale decomposition of the data. Such methods are appropriate for the analysis of non-stationary time series and are increasingly accurate at locating singularities and jumps in the data. Interested readers can find methodologies based on wavelet resolution analysis and their application in finance in econophysics publications, especially in *Physica A* (Struzik, 2001; Yalamova, 2003). So far, mainstream finance journals remain unreceptive to complexity dynamics, a critical aspect of any complex dynamic system, and as the empirical evidence of invariant laws is arguably inconclusive, they still remain loyal to the narrow view of equilibrium and efficient markets, the perspective defined by the EMH (Fama, 1970).

Our goal is not to refute the efficient market hypothesis. Instead we follow a dialectical logic. The essence of our proposed model is the recognition of the pervasiveness and value of efficient markets (Fama, 1970). But we also suggest the Hurst exponent as an indicator of the R1 tipping point from E-M trading at the *'Triple Point'* to opposite of herd trading at the *'Critical Point'*, that is, the market crash point, which Mandelbrot points out is much more frequent than what the logic of the EMH presumes (cf. Mandelbrot and Hudson, 2004). The development of this contradiction is encapsulated in dialectical logic's abstract law about the unity of opposites.

What we bring to light in this chapter is the $R1$ tipping point between efficient market trading behaviour and the beginning of bubble-build-ups by traders' interactive learning, herding, and rule-based trading, which give rise to power-law described volatility distributions. The use of the Hurst exponent as a means of identifying the tipping point where independent trading transforms into herd behaviour is critical to the instigation of the formulaic resilience engineering interventions we outline in Chapter 9.

SECTION III
Understanding How: Turning Financial Services Systems into Resilient Systems

Introduction

Erik Hollnagel and Gunilla Sundström

The focus of Section II was on the ability of a resilient system to make sense of unexpected events and to adjust what it does such that normal performance can be sustained. In addition, a resilient system must also be able to anticipate what is going to happen, both with regard to the developments of the operating environment and with regard to the impact or long-term consequences of its own performance. The implication of this for managers and regulators of Financial Services systems is a need of methods that can deal with tightly coupled and intractable systems.

A key assumption in the present section is that Financial Services systems can be actively managed and that appropriate governance processes can make the overall system more resilient. However, management and governance processes can only be effective if they can cope with both tight couplings and intractability. Two other key aspects of Financial Services systems are introduced in the present section. In Chapter 8, Grote points out that management of uncertainty is at the core of the business model of any Financial Services system. Systems in other industries (for example, nuclear power generation) generally view uncertainty as a threat to their business and operations. As a result, the overall strategy is to try to minimise uncertainty. Financial

Services systems must, however, be able to tolerate uncertainty and preferably manage it to make the system profitable. In fact, since the probability of positive outcomes is more important for Financial Services systems than the risk of failure, the ability effectively and proactively to manage uncertainty is considered a key capability of a resilient Financial Services system. This means that the system must be able rapidly to adjust its action once such potentially positive outcomes have been anticipated. One way of achieving that is to change the uncertainty management strategy. In her chapter, Grote outlines a decision process designed to help determine which uncertainty management strategy is appropriate, that is, whether to reduce, maintain or increase uncertainty in a particular situation. Uncertainty management is critical because regulation and control require a high degree of transparency and predictability, and that both of these attributes can be brought about by reducing uncertainty. Regulators therefore need to understand the uncertainty management strategy of a particular Financial Services system.

In Chapter 9, McKelvey and Yalamova introduce the idea of so-called 'resilience interventions'. These should be automatically triggered once a financial market is getting closer to the 'tipping point', that is, a point where structural system changes emerge and indicate that a crash might be imminent. The assumption is that a resilient system is able to take anticipatory action to prevent bubble-build-ups and/or lessen their negative impact. And assuming that such bubble-build-ups always will occur under certain conditions, a timely and correct intervention may lessen or avoid their negative. In Chapter 9, McKelvey and Yalamova describe an ideal scenario with technology developed to monitor market behaviour and take actions that will increase system resilience. Two categories of actions are proposed, that is, those that are based on disclosure of information from trusted sources and those that are designed to target 'reckless behaviour' of individuals and/or individual Financial Services firms.

Finally, in Chapter 10, Sundström and Hollnagel suggest that regulatory changes really only can have an impact if there is a transparent and shared view of the Financial Services system being monitored. One way of establishing that is a functional modelling approach that focuses on the tightly coupled system functions rather than complex system structures. In addition to a shared view of the system, regulatory entities must also have a shared view of early warning signs (leading indicators) and the key performance indicators for a system's health. Finally, recognising that a resilient system has four key capabilities, that is, the ability to respond, to monitor, to learn, and to anticipate, it is concluded that Financial Services systems governance and control processes must be designed to determine the presence of these four capabilities.

Chapter 8
Balancing Different Modes of Uncertainty Management in the Financial Services Industry

Gudela Grote

Organisations will often seek to reduce uncertainties in order to ease goal achievement based on predefined plans and programmes. Sometimes, though, they will deliberately increase uncertainties as a way to gain competitive advantage, as has been the case in recent years in the Financial Services industry. The chapter will discuss different modes of uncertainty management, especially minimising versus coping with uncertainty and their consequences for the functioning of organisations. Achieving an appropriate balance between stability and flexibility and a match between power, control and accountability will be proposed as important criteria for making decisions on the management of uncertainty. Decisions have to systematically consider the costs and benefits of reducing, maintaining, or increasing uncertainties as well as changing the belief systems in the organisation that influence the assessment of these costs and benefits. Current discussions on reducing uncertainty in the Financial Services industry will be reflected upon on the basis of the suggested approach.

The importance of uncertainty for organisational functioning has long been recognised: 'Uncertainty appears as the fundamental problem for complex organisations, and coping with uncertainty as the essence of the administrative process' (Thompson, 1967: 159). Organisations differ substantially, though, in terms of the uncertainties they are faced with and the requirements for turning uncertainties into business opportunities. Two examples are investment banking and running a nuclear power plant. While investment banking capitalises on fast and risky decision-making in highly uncertain financial markets, nuclear power plants are concerned with shielding their processes from uncertainty and maintaining routine operation for as long as possible. Financial Services firms need certain structures and standards to ensure responsible handling of investment risks, whereas nuclear power plants face requirements for handling non-routine situations and for innovation in response to technological and regulatory developments. Analysis of internal and external uncertainties and decisions on how to manage them on a strategic and operational level are key to successful performance in both cases, even though the decisions taken will differ substantially.

One fundamental difference between the Financial Services industry and other high-risk industries like process industry or transportation renders the uncertainty management perspective particularly interesting. While in these other industries, uncertainty is mainly regarded as a threat, in Finance Services uncertainty lies at the heart of the business model. 'A financial market requires the certainty that the uncertainty will continue' (Brügger, 2000: 252). This is reflected in the belief systems in the respective organisations and the attitudes and values of the individuals working there. Uncertainty is considered as an opportunity and individual and organisational capabilities are tailored to continuous risk-taking. Coping with uncertainty is the dominant strategy, while in other high-risk industries minimising uncertainties is preferred. The starting point for approaching an appropriate balance between stability and flexibility in Financial Services is therefore opposite to most other industries, which has to be kept in mind when strategies for risk management are to be compared or even generalised between industries.

In the following the different strategies for managing uncertainty will be explored and a decision process suggested for deciding on the most appropriate strategy. Before getting further into the topic, a word is needed on the chosen focus of uncertainty and its management as compared to risk and management of risk. Risk is commonly defined in terms of possible loss or damage, but in finance it is understood more broadly as the possibility of deviation from a desired outcome (Gallati, 2003). Uncertainty is considered to be the more generic concept, which can easily be linked to risk in its most basic form as an uncertain event or in more specific definitions of risk such as the product of probability and damage. Uncertainty in terms of insufficient or ambiguous knowledge about cause-and-effect relationships can be regarded as the 'neutral' source of risk. Uncertainty implies that predictability and transparency as crucial prerequisites of governance and control are reduced, which leads to insufficient means to either avoiding damages or realising opportunities. Power (2008) postulates that uncertainty is transformed into risk when it becomes an object of management. When uncertainties are managed well, a basic prerequisite for good risk management is established.

With the proposed focus on uncertainty it also becomes easier to appreciate the relevance of both decreasing and increasing risk. Many business operations capitalise on exploring new territory, which by definition means increasing uncertainty and risk, even though this is often not made explicit due to the generally negative connotations of risk. Using Knight's distinction between risk and uncertainty with the one being grounded in calculable probabilities and the other being incalculable, Beunza and Stark (2005) argue that entrepreneurship is not about risk-taking, but about exploiting uncertainties. Even when uncertainties are turned into calculable risks based on some financial theory or technology, the uncertainty remains which formula or algorithm to use. 'Entrepreneurship is the ability to keep multiple evaluative principles in play and to exploit the resulting ambiguity' (Beunza and Stark, 2005: 91).

Risk management entails the identification and evaluation of risks, decisions on measures for handling the risks, and risk communication (Renn, 2008). Correspondingly, uncertainty management concerns the identification and evaluation of uncertainty, decisions on measures for handling uncertainty, and uncertainty communication. All of these elements will be touched upon in the following, but the focus will be on the decision process necessary for determining measures for handling uncertainty. The risk management literature usually discusses four ways of handling risks: reduction, retention, avoidance, and transfer (for example, Renn, 2008). This leaves out a fifth option, which may be of equal importance that is, deliberately increasing risks. The significance of this fifth option is particularly obvious in the Financial Services industry (MacKenzie, 2006). Therefore, both the reduction of risk, which in a more general sense also comprises avoidance and transfer of risk, and retention of risk, which is expanded to also include increasing risk, will be considered. Translated into the chosen focus on uncertainty, this means that reducing, maintaining, and increasing uncertainty are viable options to be considered in the suggested decision process.

In the next sections some generic considerations about managing uncertainty are laid out, and then applied to the Financial Services industry.

Minimising Uncertainties versus Coping with Uncertainties

As a starting point for making strategic decisions on how an organisation should approach uncertainties, minimising uncertainties versus coping with uncertainties can be contrasted (Grote, 2009). The major approaches to system design at the turn of the twentieth century (Taylor, 1911; Weber, 1947) were built on the assumption that organisations were closed systems protected from external uncertainties. Concerning internal uncertainties, the assumption was that they should be minimised by minute planning and continuous monitoring of the execution of these plans, providing minimal degrees of freedom to the people in charge of carrying out the plans and taking any deviation from the plans as signs for the necessity of even more planning and monitoring (see Figure 8.1). Accordingly, the basic control mode is that of feedforward control. The Fordist production lines are a prime example of the minimising uncertainties approach. They were tailored to mass production of standard products, that is: Model T in black.

With the acknowledgement of the open system nature of organisations this approach continues to be followed and has even gained in fervour in order to keep systems under control: if the organisation's environments become more and more complex – or rather the complexity is more and more acknowledged – then more effort has to be put into reducing the uncertainties

Minimising uncertainties	Coping with uncertainties
• Complex, central planning systems	• Planning as resource for situated action
• Reducing operative degrees of freedom through procedures and automation	• Maximising operative degrees of freedom through complete tasks and lateral cooperation
• Disturbances as to be avoided symptoms of inefficient system design	• Disturbances as opportunity for use and development of competencies and for system change

▼   ▼

Dependence/feedforward control   *Local autonomy/feedback control*

▲ *Balance through loose coupling* ▲

Intrinsic motivation

Participation in rule-making

Flexible changes between organisational modes

Culture as basis for coordination/integration

Figure 8.1   Basic principles of uncertainty management
Source: Adapted from Grote, 2004b.

connected with the complexity. As the minimising uncertainties approach promises maximum control, it is still the favoured approach in many organisations.

Another approach which has been promoted by organisation theorists and work scientists for several decades is to enable all members of an organisation to cope with uncertainties locally and to rely on feedback control (for example, Perrow, 1967; Weick, Sutcliffe and Obstfeld, 1999; see Figure 8.1). From this perspective, planning is understood primarily as a resource for situated action (Suchman, 1987), not as a blueprint for centrally determined and monitored action. Local actors need to be given as many degrees of freedom as possible, achieving concerted action mainly through lateral, task-induced coordination. Disturbances are also regarded as opportunities for use and expansion of individual competencies and for organisational innovation and change. Cherns' (1987) principles of socio-technical design provide a good summary of the core ideas of this approach, especially the principles of minimal critical specification for work processes and task allocation, of role breadth to ensure multifunctional expertise, and of controlling variances at their source.

Balancing Stability and Flexibility in Organisations

Much of the earlier literature in organisation theory was aimed at developing contingency models for deciding between these two approaches in light of the types and amounts of uncertainty a particular organisation is faced with (for example, Argote, 1982; Burns and Stalker, 1961; Van de Ven et al., 1976; for a comprehensive review see Wall et al., 2002). The most basic understanding of these contingencies is that minimising uncertainties only works when the overall level of uncertainties an organisation must confront is low. With higher levels of uncertainties, any attempt to design them out of the system will fail and therefore the system has to be enabled to cope with uncertainties locally.

More recently, research has been concerned with showing the need and also the possibilities for overcoming the dichotomy between the two approaches to managing uncertainty. As early as 1976, Weick argued that most organisations aim to achieve what he called loose coupling that is the concurrence of autonomy and dependence and thereby also a mix of coping with and minimising uncertainty. Elements of loose coupling are intrinsic motivation, which promotes using autonomy in line with superordinate goals, participation in rule-making, mechanisms that support swift shifts between the two modes of handling uncertainty, and finally culture as a 'soft' form of centralisation through values and basic assumptions (see Figure 8.1).

In 1991, March wrote a very influential article, approaching the same issue from the perspective of learning in organisations. He argued that a balance is needed between exploration of new possibilities, concerned with search, variation, experimentation and risk taking, and exploitation of old certainties in terms of refinement, implementation, and efficiency. In the competition for scarce resources in organisations, exploitation tends to win because benefits are more visible and short-term (Benner and Tushman, 2003). As an example, March discusses the socialisation of newcomers into organisations, pointing to the usual attempts in organisations to ensure fast learning of organisational routines in order to quickly reach efficient performance at the expense of the organisation learning from the different viewpoints and prior experience of the new employee.

March's work has motivated much research into achieving a balance between exploitation and exploration and thereby also, between stability and flexibility (Gupta et al., 2006). In strategic management, Tushman and Reilly (1996) have coined the term ambidexterity to indicate the concurrent demands of stability and flexibility on organisations. It has also been pointed out that this balance requires new management styles that can handle the apparent contradictions between coping with and minimising uncertainty (Smith and Tushman, 2005), for instance regarding the demand for both empowering and directive leadership.

Even in the classic high-risk industries it is now acknowledged that organisations need both the stability created by minimising uncertainty and the flexibility achieved by coping with uncertainty. The concepts of high-reliability organisation (for example, Weick, Sutcliffe and Obstfeld, 1999) and of resilience engineering (Hollnagel et al., 2006) are prominent examples of this shift in thinking. In order to fulfil requirements of concurrent stability and flexibility that neither minimising nor coping with uncertainty can accomplish alone, a more detailed consideration of the usefulness of reducing, maintaining and increasing certain types of uncertainties is required. In the next section, the different steps to be taken in such decisions are described.

Deciding on a Management Strategy for Handling Uncertainties

As a starting point we make the assumption that the overall objective in individual and organisational decision-making is to gain and maintain control in order to achieve desired goals. For decision-makers in organisations, uncertainty may in itself induce strong perceptions of threat beyond the actual threats of economic loss (Argote et al., 1989), and the frequent first reaction is therefore to try to reduce them. The Financial Services industry is the prime example of the not so common opposite tendency to seek uncertainty and promote risk-taking as a competitive advantage. The decision process described in the following aims to achieve more balanced ways of handling uncertainty in a given organisation by looking at advantages and disadvantages of both reducing and maintaining or even increasing uncertainty.

As a preparatory step, an uncertainty map needs to be drawn up that includes as many of the relevant internal and external uncertainties as possible. For these uncertainties, an assessment will be performed along four steps (for more detail see Grote, 2009). First, the costs and benefits involved in reducing, maintaining or even increasing uncertainty are considered. This assessment is then complemented by a reflection of belief systems developed, enacted and continuously recreated in the company concerned in order to reduce fundamental uncertainties about appropriate management practices. Finally, recommendations derived from these considerations need to be evaluated and possibly revised. The suggested steps in the decision process are the following:

- Analyse costs and benefits of reducing uncertainty.
- Analyse costs and benefits of maintaining or increasing uncertainty.
- Explore belief systems in the organisation related to managing uncertainties.
- Discuss anticipated effects of the recommendations derived in steps 1 to 3 and repeat any step as necessary to come to a final decision.

Step 1 of the Decision Process

The first step conforms to most classic treatments of uncertainty management and entails an analysis of costs and benefits of reducing uncertainty. Uncertainties stem from insufficient transparency and predictability of situations due to lacking or ambiguous information and from unclear options for taking action in these situations. Transparency and predictability can be created either by gathering more information or by eliminating the causes of opaqueness and unpredictability. Especially the latter is usually closely tied to the use of power in order to force other actors to disclose their plans, to agree to binding arrangements or to have uncertainty transferred to them. The more uncertainties there are, the more costly any reduction strategy becomes. Moreover, while the benefits of reducing uncertainty are quite obvious because control is increased, the costs are partially hidden. In particular, the loss of flexibility associated with uncertainty reduction is not always sufficiently taken into account.

The general recommendation to be drawn is that reduction and control of uncertainties is frequently a desirable strategy, which, however, focuses perception on the expected. Thereby, threats as well as opportunities may be overlooked. Also, these strategies can be very costly due to resources spent on measurement and control of internal and external processes. Finally, they do require some degree of predictability and in the case of the power response, a very strong position vis-à-vis other actors.

Step 2 of the Decision Process

The second step looks at the costs and benefits of the opposite strategy of maintaining or increasing uncertainty. As a first general consideration, it is important to note that more information – usually discussed as a means to reduce uncertainty – may actually create new uncertainties if the information concerns events with unknown probabilities or allows for different interpretations (Becker, 2004). To date there is little empirical research on the deliberate increasing of uncertainties due to the dominant view of uncertainties as inevitable, but largely unwanted. The exception is the literature on the Financial Services industry which we will turn to later. Even in the innovation literature, uncertainties tend to be acknowledged only to the extent that they are an unavoidable side effect of discovery. Suggestions have even been made on how more routinised processes can be introduced to increase the predictability of innovation (for example, Brown and Eisenhardt, 1995; Nelson and Winter, 1982).

What is more frequently discussed is maintaining uncertainty and responding to this uncertainty by increased internal flexibility. 'Unlike control and cooperation strategies which attempt to increase the predictability of important environmental contingencies, flexibility responses increase internal responsiveness while leaving the predictability of external factors unchanged' (Miller, 1992: 324). A classic example of this strategy is diversification, be it with respect to financial products, markets or suppliers, which reduces resource dependence.

Iterating Steps 1 and 2

In order to find a proper balance between stability and flexibility not all uncertainties should be handled in the same way. The basic assumption underlying the first two steps is that reducing uncertainty usually increases stability, while maintaining or increasing uncertainty supports flexibility. However, aiming to reduce uncertainty that would better be maintained can actually destabilise the system. For instance, if the sequence of work processes is fixed with no decision latitude given to people at the operative level to adapt it in response to local disturbances, this may severely hamper the workflow in the affected unit and beyond.

Overall, there seems to be comparatively little to recommend increasing uncertainties, unless radical innovation in highly volatile environments is the organisation's main objective (for example, Eisenhardt and Martin, 2000), or unless uncertainty is itself key to business operations, as in Financial Services. The costs of acknowledging limited control, of increasing the variety in possible responses to external contingencies, and of building buffers are seen to easily outweigh the benefits of flexibility and responsiveness. However, this view is often skewed by overly optimistic beliefs about one's own influence and power. When assumptions about available options for reducing uncertainties turn out to be false, the cost-benefit calculation for maintaining or even increasing uncertainties may start to look more favourable. On the other hand, there may be unrealistic beliefs about the ability to stay in control even with very high levels of uncertainty, as the recent financial crisis has aptly demonstrated.

Besides the rational approach to understanding and managing uncertainties, it is therefore important to be aware of the belief systems that individuals and groups develop with respect to uncertainties and their control. Different attitudes, norms and values associated with uncertainty may turn out to be more powerful determinants of decisions than the rational considerations presented so far. These belief systems and their influence on decisions regarding the management of uncertainty are the primary focus of the third step.

Step 3 of the Decision Process

The third step involves switching perspective from rational accounts of cost-benefit analyses to one of sense-making and enactment (Weick, 1995). This perspective holds that, for decision-making, perceptions of uncertainty are more relevant than objective accounts, and that these perceptions and the actions derived from them are embedded in and shaped by decision-makers' belief systems. Either minimising or coping with uncertainty may be the preferred way of managing uncertainty based on beliefs about control and trust. These beliefs will make certain responses to uncertainty more appealing than others, even though these responses may be quite ineffective due to the particular conditions. Two examples regarding the effects of values and basic assumptions as part of professional cultures may serve to illustrate the importance of belief systems in decision-making.

Weitz and Shanhav (2000) postulated that engineers used their success in handling technical uncertainties to expand their professional domain to include the reduction and elimination of organisational uncertainties as well. How brittle this approach is, is shown by Feldman's (2004) analysis of two major NASA accidents: the explosions of the shuttles Challenger and Columbia. Feldman traces some of the faulty decision-making involved in these tragedies back to an over-confidence in quantitative data combined with neglect of non-quantifiable data. As an underlying cause, he sees the culture of objectivity at NASA, a culture he considers typical for an engineering organisation. 'They (the NASA engineers) were not able to quantitatively prove flight was unsafe, so in this culture it became easy for management to claim it was safe … Under conditions of uncertainty, cultures dominated by the belief in … objectivity must be silent. This silence makes these cultures vulnerable to power and manipulation' (Feldman, 2004: 708).

More directly related to the current focus on the Financial Services industry is an article by Ferraro and colleagues (2005) in which they investigate belief systems of economists and how they sustain certain theoretical assumptions and organisational practices. In particular, they discuss the core assumption of humans being motivated mainly by economic self-interest. They argue that one effect of training in economics is to strengthen beliefs in the pervasiveness, appropriateness, and desirability of self-interested behaviour to a point where these beliefs do not only influence individuals' behaviour, but also permeate economic theory. The emphasis placed on market mechanisms for handling conflicts of interest and the importance given to external incentives in understanding and influencing behaviour are examples of that. MacKenzie's (2006) analysis of the effects of option pricing theory provides a striking illustration of how beliefs embedded in financial theory and handed down to those making the actual decisions will finally shape reality to conform to theory.

Step 4 of the Decision Process

In the fourth step, an overall evaluation of the chosen and potentially readjusted strategy for handling uncertainties is carried out. The aim is to achieve a balance of stability and flexibility fitted to the particular needs of the organisation. Such an evaluation is, of course, difficult to make. Instead of aiming for some absolute judgement, often it will be more helpful to define probing strategies for continuous re-evaluations of the achieved balance. In this step, the recursive nature of the whole process is also considered. The outcome of step 4 may be that no satisfactory way forward can be defined leading the discussion back to the previous steps. Also, during any of the other steps it may become apparent that previous steps need to be reworked.

Managing Uncertainties in the Financial Services Industry

As was stated before, contrary to most other industries where the first choice is usually to reduce uncertainty, in Financial Services the main thrust during the last three decades has been to thrive on increasing uncertainty. There have been warning voices and some attempts at reducing uncertainty through increased regulation like the Basel II accords over the last several years (Power, 2007). Only after the most recent financial crisis, however, has it become more widely acknowledge that uncertainty must be managed differently in the future. The collapse of Barings Bank brought about to a large extent by one 'rogue trader' in

1995 turned people's attention to so-called operational risk and the need to increase oversight over the internal workings of banks. The financial crisis has forced everybody to acknowledge that not only do banks not handle operational risk as well as they should, but that risks fundamental to the workings of the Financial Services industry as a whole have gotten out of hand, that is, credit risk and market risk. Much of the current discussion revolves around different kinds of regulations that will reduce uncertainty in these three risk domains, be it through codes of conduct and stronger surveillance or by cutting back certain types of financial operations. Another fiercely conducted debate concerns necessary changes in the incentives for individual and organisational risk-taking behaviour, ranging from reworking bonus systems to abandoning the *too big to fail* principle.

Obviously, this chapter cannot provide straight answers to these fundamental issues, but the perspective taken here on choosing between different strategies for managing uncertainty is hoped to be helpful to those who will have to find these answers. I will use two examples to illustrate the suggested approach, namely the use of quantitative models in trading decisions and rules governing foreign exchange trading. In the final section I will discuss as an overarching principle the importance to manage uncertainties in a way that provides the best possible match between power, control and accountability.

Managing Uncertainty through Quantitative Models for Trading Decisions

For a long time, information systems have been used to support financial decision-making. Expert systems for evaluating the credit worthiness of customers are one example, another are support systems for active portfolio management in investment banking that are taken as an example here.

Svetlova (2008) studied different strategies used by portfolio managers for making trading decisions. Specifically, she compared qualitative heuristics to quantitative, computer-based strategies. On the one hand, portfolio managers use a wide variety of heuristics, that is, simplifying rules based on a selection and prioritisation of factors which are closely observed and acted upon, such as job markets in particular countries, bonds with particular running times, or inflation rates. Also, information sources are prioritised: some trust company reports, others focus on personal informants or on assessments by brokers and analysts. Information is combined using self-made simple tools like spreadsheets. Individual strategies are encouraged by management in order to take maximum advantage of individual competence. The multitude of decision strategies, while aiming at individually coping with uncertainty, increases uncertainty overall, which creates new opportunities for competitors to outperform each other in coping with the uncertainty.

On the other hand, formal computer-based models are employed, which are based on statistical analysis of historical data. These models can be used to inform human decision-making, but they are increasingly used also as expert systems that autonomously execute trading decisions. However, extensive and uniform use of similar formal models may create 'computer herding' that endangers market dynamics as all actors want to either buy or sell (Tett and Gangahar, 2007). In the end decision-making has to remain with humans who must compensate for shortcomings in the formal models and make the final judgement on the decision strategy or combinations of strategies to be employed.

From the perspective of uncertainty management, it is obvious that decision-makers are quite aware of the need to balance efforts aimed at reducing, maintaining and increasing uncertainty in order to keep the basic dynamics of financial markets intact. While formal models, and to a lesser degree qualitative heuristics, are used to reduce uncertainty and to cope with remaining uncertainties, there is also an understanding that streamlining these methods in line with a minimising uncertainty approach may prove harmful. Compared to other industries, there is a clear awareness that increasing uncertainty is beneficial for business as long as there is requisite flexibility to handle it. The tools available to support that requisite flexibility have to be employed very carefully in order for decision-makers not to get trapped by the fact that the tools have been shaped by their own theories and beliefs. Social interaction during decision-making has been shown to be crucial to balance different perspectives embedded in the different tools and models (Beunza and Stark, 2004).

Managing Uncertainty in Foreign Exchange Trading

Foreign exchange trading only started when fixed exchange rates as set by the Bretton Woods agreement were given up in the late 1970s. This created a new market that has grown to be the largest and fastest moving financial market. Within foreign exchange trading, interbank currency spot trading is probably the fastest operation. Decisions in spot trading must be made in a few minutes, or even faster, and are embedded in fast, highly structured and truncated sequences of electronically mediated communication, so-called trading conversations (Knorr Cetina and Brügger, 2002). Individual traders are highly autonomous in their decision-making and act as market makers more than as agents of an organisation. Limits set by their employing organisation only concern maximum losses and overall volume of currencies traded. At the same time, trading performance is very closely monitored, by the traders themselves as well as by their supervisors. The trader exposes him- or herself with every decision made and the ensuing extreme tension finds an outlet 'in a vocabulary that resounds with the emotions of perceived violence and attack' (Knorr Cetina and Brügger, 2002: 939).

The following example conveys the structural elements and pressures involved in spot trading very well (Knorr Cetina and Brügger, 2002: 938):

1	FROM GB5I <Name of Bank> MILAN *1135GMT 251196 */3447
2	Our terminal: GB1Z Our user: <Name of Spot Dealer>
3	SPOT CHF 5
4	#<InSD> 6364
5	FROM <IS>
6	# #INTERRUPT#
7	#
8	#INTERRUPT#
9	
10	# #INTERRUPT#
11	# HALLOOOOOO THIS IS SPOT AND NOT FWDS OK?/
12	YES MATE SRY CUST.
13	#
14	#INTERRUPT#
15	MY RISK PSE
16	#
17	NWOSPE
18	# 6263

19   SELL
20–27   <confirmation and closing sequence>

In this conversation, the trader, who was 'king of the floor' – in terms of earnings for the bank, reputation among traders, and importance of the currency traded – offers a narrow price range (line 4). The trader is then kept waiting by the caller, who does not respond and tries to snatch the turn back from him (line 8) while the trader also holds on to it or also interrupts (lines 6 and 10). This back and forth ends when the Zurich trader interjects an angry reprimand (HALLOOOOOO THIS IS SPOT AND NOT FWDS OK?), thereby reminding the caller that, in spot trading, rather than situations in which long-term instruments are traded, a response must come forth immediately. The caller, accepting this, apologises by asserting that he himself was kept waiting by a customer who had not reached a decision (line 12 means 'Yes, mate; sorry, customer'). The caller interrupts the flow once more and then offers MY RISK PL[EA]SE – meaning that he considers the Zurich trader no longer committed to the price. The caller accordingly repeats the price question when ready (NWOPSE, a misspelled and abbreviated form of 'Now please', [line 17]), receives a new price offer with an equally narrow spread (6263), and agrees to the deal (SELL)'.

The high level of individual decision-latitude is bound by informal rules of conduct as mechanisms of social governance: a trader's initiating question has to be uttered in a neutral way that does not disclose the caller's intent regarding buying or selling. The responding trader has to answer by indicating a price for both options. Traders are committed to the price that they indicate, as long as the caller responds within a timeframe of about two seconds and as long as the trader has not regained the turn in the conversation by an interrupt key in order to invalidate the price. Whatever price is offered is non-negotiable within the given trading interaction. The code of conduct creates institutionalised expectations against which individual trading behaviour and trading relationships are assessed. These informal rules regulate behaviour for which very few legal sanctions exist and for which legal sanctions are also considered inefficient by the actors involved.

Interpreting the described interaction patterns from the perspective of managing uncertainty, it is very obvious that the work system is set up to handle very high amounts of uncertainty in a very flexible fashion. Uncertainty is managed individually based on high levels of very specific expertise, possibly using other traders in the same and in other markets as resources (Beunza and Stark, 2004). The main stabilising element is the informal code of conduct that constrains trading interactions and thereby makes them more predictable and controllable.

Obviously, the combination of high stakes, individual autonomy and time pressure easily lends itself to undue risk-taking. In order to keep risk-taking at bay, some simple rules have been in place for a long time, such as the two-week vacation rule for traders that is intended to interrupt spirals of increasingly risky decisions to make up for prior losses. Any further regulation aiming at the immediate trading behaviour will have to be carefully evaluated against the existing informal code of conduct and how that may be negatively affected by introducing more formal rules. Traders themselves will argue, as mentioned already, that legal sanctions are ineffective to control trading behaviour. Whether this claim is true would have to be investigated more systematically, also taking into account that traders would probably be very critical regarding any measure reducing their autonomy. At the same time it may very well be that certain rules could actually help them to better cope with the uncertainties and pressures in their jobs. The tools described in the previous example could also be regarded as rules in themselves that guide behaviour without fully constraining it.

On a much more fundamental level, the question has been raised whether exchange rates should be fixed again so as to eliminate all uncertainty and risk by stopping foreign exchange trading altogether. The follow-up question is how such a drastic measure might negatively affect the Financial Services industry and the wider economy. Again individual and organisational belief systems will come in strongly in such an evaluation, especially those very basic beliefs pervasive in economics regarding markets as the most effective form of governance (Ferraro et al., 2005).

Organisational Governance: Matching Power, Control and Accountability

The management of uncertainty can be understood within a general framework of control, where control refers to the ability of an individual or a system to influence situations for achieving certain goals. One can add spice to this rather neutral description by arguing that the ability to handle uncertainties also defines power in organisations (Hickson et al., 1971; Marris, 1996; Clegg et al., 2006). How the power to control uncertainties is distributed needs to be evaluated against the effects on individual and collective accountability (Monks and Minnow, 1991). Whoever is in control also is to be held accountable, thereby constraining the abuse of power.

While often in the design of socio-technical systems the main concern is that individuals are held accountable for outcomes of work processes over which they have insufficient control (Grote, 2009), in the Financial Services industry the opposite problem is discussed: people that are in control – in as much as control is possible in these highly uncertain work processes – are not, or at best only partially, held accountable for the results of their actions. In order to improve governance in these organisations, power relationships will have to be readjusted in a way that permits to better link control of uncertainties with accountability for outcomes.

The Impact of Increasing Regulation

Currently, introducing more rules and regulations is discussed as the main approach to achieve this readjustment. One general concern in drawing up rules is to carefully assess specific action constraints to be imposed in order not to restrict individual autonomy to a degree that is harmful to controlling necessary uncertainties and as a consequence also to ascribing individual accountability (Grote et al., 2009). Rules can set certain goals to be achieved (goal rules), they may also define the way in which decisions about a course of action must be arrived at (process rules), or they can prescribe concrete actions (action rules) (Hale and Swuste, 1998). Overall, action rules are most appropriate within a minimising uncertainty strategy, while process and goal rules are supportive of a coping with uncertainty approach. Following, some more specific considerations regarding the appropriateness of different rules types are provided:

- Goal rules may be very useful in pointing out desirable outcomes and priorities and thereby providing an overall orientation for the actor. The much debated 'uptick' rule which allows short selling only for stocks that have just increased in price, may be understood as a goal rule because it signals to all market participants that making money out of somebody else's misfortune is undesirable (MacKenzie, 2006).
- Process rules are particularly well suited to the design of stable but flexible work processes. These rules provide guidance through heuristics for information search, decision-making, and problem-solving. Such rules can help financial decision-making, for instance, by supporting the ability to handle the paradox created by concurrent requirements for prudence and risk seeking, discipline and flexibility, and reasoning and intuition (Brügger, 2000).
- Action rules are for many the 'real' rules or standard operating procedures in the sense of prescribing concrete actions. Highly prescriptive and detailed action rules are the rules that people usually have in mind when they criticise the rigidity of standardised work processes, because their main strength is also their main weakness: action rules provide very detailed guidance for actors, which reduces the demands on knowledge, expertise and resources spent on action regulation and coordination. If the rule fits the situation perfectly, this will usually result in 'perfect' action, if it does not, chances of actors realising that and adapting their behaviour accordingly are severely reduced. This problem was discussed earlier already with respect to the rules embedded in quantitative models for trading decisions (Beunza and Stark, 2004; Svetlova, 2008).

Once appropriate control is established by providing a good fit between constraints on actions and the uncertainty these actions have to accommodate, the next step for building good governance is to link that control to accountability. Rules put in place need to be scrutinised with respect to them leading to a delegation of responsibility to rule-makers and rule-inspectors instead of fostering personal accountability. This is most likely to happen when personal initiative is constrained so much that people feel they are acting out somebody else's will. As Roberts (2001) argues, proper governance can be achieved best when hierarchy- and power-bound accountability is complemented by what he calls socialising forms of accountability. These are established by close contacts between people in a relative absence of formal power differentials and depend upon individuals' willingness to risk speaking up. The trading conversation described earlier and the rules of conduct enacted in them are a good example of these socialising forms of accountability. However, they obviously could not prevent rogue trading, which is related less to the code of conduct as such but to the incentive systems fostering excessive risk-taking. This brings me to the last point, which is the necessity to gear incentives towards rewarding responsible behaviour.

Incentives for Responsible Behaviour

Rewarding responsible behaviour will take out some of the challenge in jobs in the Financial Services industry. Instead of being in a constant hype at the boundaries of risk-taking turning into gambling, more conservative decisions are required. How this may in fact change the profile of people being attracted to working in Financial Services one can guess at based on a study by Philippon and Reshef (2009). They found that qualification and pay in banking was higher before 1930 and after 1980 compared to other industrial sectors. This was interpreted in light of more regulation introduced as a consequence of the financial collapse in the early 1930s and deregulation undertaken in the early 1980s. Higher levels of regulation reduced the complexity of jobs due to fewer uncertainties that needed to be handled. These less complex jobs attracted less qualified people and had lower pay attached to them. This study seems to confirm worries about losing very competent employees when more regulation regarding both financial operations and compensation systems is introduced. But one may also argue that it is perfectly in order to have less qualified people do simpler jobs for less pay if this is what turns banks into more responsibly run enterprises.

An interesting case in this respect is the sale of Citigroup's energy trading unit, which was presumably undertaken to get rid of bonus obligations towards the main trader that cannot be fulfilled under the new legislation (*Herald Tribune*, 16 October 2009). While some consider this decision and the legislation that caused it a mistake because Philbro was very profitable and the trader was a particularly successful one, others applauded it as it freed the bank of particularly high risks.

Trust and the Limits of Control

The very final word is on trust. However well uncertainty is managed and the most appropriate match between power, control and accountability achieved, there are limits to control, which have to be bridged by trust both in people and in systems (Luhmann, 1979). Organisations have the power and presumably also the knowledge to make sensible decisions on risk involved in business operations. This creates the rightful expectation that they can also be held accountable for the decisions they take. However, the ensuing concern of living up to this expectation may lead risk experts to frame their judgements more in terms of reducing their personal, legal and reputational risks than in terms of providing honest and trustworthy assessments of the risks at hand (Power, 2004). This may create the paradox that focusing too narrowly on risk management becomes itself a risk. In order to avoid this problem, Power argues for a new politics of uncertainty that 'would not seek to assuage public anxiety and concerns with images and rhetorics of manageability and control, and would challenge assumptions that all risk is manageable. Public understandings of expert fallibility would be a basis for trust in them, rather than its opposite' (Power, 2004: 63).

While this new politics of uncertainty is very useful to promote open dialogue about risk, it clearly also has the downside that decision-makers may be encouraged to disclaim their contribution to failures, as has happened repeatedly in the Financial Services industry. In order to live up to rightful expectations of responsible decision-making and to promote supervising structures acting as 'guardians of trust' (Shapiro, 1987), decisions have to be based on explicit scenarios that demonstrate how adequate coping with uncertainty and risk can be achieved. However, these scenarios must also include the acknowledgement of limits of controllability and the definition of accountability for business processes within and outside these limits. In view of responsibly handling the particularly high uncertainty and risk involved in Financial Services, MacKenzie (2006) has called for broad conversations on the design of financial markets in order to help build and maintain Financial Services systems that may serve the interests of all. The suggested process for deciding on the most appropriate strategies for managing uncertainty is hoped to promote such conversations in organisations and beyond.

Chapter 9

Financial Resilience Engineering: Toward Automatic Action Formulas against Risk and Reckless Endangerment

Bill McKelvey and Rossitsa Yalamova

Chapter 4 introduced Econophysics as a way to break loose from existing financial persuasions, resistances, and Financial Engineering (FE) methods. In particular, an econophysics perspective helps us to move away from assumptions associated with Louis Bachelier's (1914) random-walk model of stock price behaviour and Gene Fama's (1970) widely accepted view that markets always behave according to the efficient market hypothesis (EMH). Ideally, our goal should be to create financial markets that are resilient, hence less prone to crashes. For example, interventions to prevent bubble-build-ups and/or lessen their negative impact would be highly desirable. In the following we will refer to these types of interventions as resilience interventions. Leveraging econophysics, we identify conditions leading to herding behaviour (Hirshleifer and Teoh, 2003), which begins with *tiny initiating events* (Holland, 1995, 2002) and sometimes results in extreme outcomes such as trading volatility, bubble-build-ups and crashes. In Chapter 5 we illustrated how numerous scalability spirals occurred in the 35 years between the invention of the Black-Scholes options-pricing model in 1972 and the 2007 liquidity crisis and the following Great Recession.

In Chapter 7 we built on a recent idea in the world of econophysics, that is, to use the volatility autocorrelation function, Hurst exponent, and power laws (PLs) to identify herding in financial market behaviour. In particular, we argued that these techniques can be used to identify the so-called tipping point, that is, the point in which market behaviour transitions from predominantly random trader behaviour to predominantly herding behaviour. As pointed out as early as 1914 by Henri Poincaré; herding sets off volatility bubble-build-ups that sometimes extend, if not stopped, to the so-called *Critical Point*, that is, a market crash. Our methods introduced in Chapter 7 leverage work by Didier Sornette and colleagues aimed at identifying the tipping point (Yan, Woodard and Sornette, 2010; Chapter 6). Our econophysics developments in Chapter 7 provide the basis for identifying tipping points; in the present chapter we assume that we have this capability and thus focus on how resilience interventions could be automatically triggered. Our view is that the engineering of such intervention mechanisms is a critical element for how financial markets can be made more resilient. In fact resilience interventions that are imposed automatically rather than depending on potential actions by lobby-influenced politicians and government regulators, are a critical element of resilient Financial Services systems.

Internal versus External Shocks

Empirical evidence indicates that the volatility signatures of system perturbations that start within a trading system (endogenous shocks) and system disturbances that are triggered by external an event such as a decision by a national government to freeze its currency (exogenous shocks) have different relaxation times. Relaxation time is the time it takes a stock-trading system to transition back to the *Triple Point*, where greed, risk, and noise are balanced, that is, a state associated with EMH-style trading behaviours. Based on a number of recent empirical studies (Sornette et al., 2002, Cajueiro et al., 2009, Zunino et al., 2009), we suggest several measures to prevent self-organisation in a market from bubbling up to the Critical Point via herding, connectivity and imitation. If our proposed resilience interventions are to minimise the longer-term effects of market crashes – such as the current *Great Recession* – which results from slower, self-organising *endogenous* process dynamics – it is critical that we create resilience against endogenous processes rather than attempt to prevent anomalies foreign to the US stock-trading system, for example – such as the Asian meltdown of 1987, the Russian bond default of 1998, the dot.com crash of 2000, or the 2009 meltdown in Dubai (2009) the 2010 Greek and Irish debt crises.

Relaxation time characterises the way a dynamical system returns to equilibrium after experiencing significant deviations. It may be the time-dependent response of a system to external stimuli, or the collapse of a self-organising system after reaching an unsustainable phase. In fact, once a system reaches the Critical Point the relaxation time (that is, the time needed for the market to retreat back from the top of the bubble to EMH-style trading at the Triple Point) is much longer because the collapse of a self-organising system creates larger volatility bursts over a longer period of time. Sornette et al. (2002) show that the rate of the relaxation pattern of *endogenous* shocks via self-organising *after* a system reaches the Critical Point takes more time than relaxation after *exogenous* shocks set off by external events, even if they occur *before* the system reaches the Critical Point.

The difference in impact between exogenous and endogenous shocks is due to the way that volatility relaxes to its EMH-based average value. Sornette et al. (2002) illustrate that endogenous shocks in the form of small perturbations, for example, the impact of brief news reports (that is, tiny initiating events) on some traders can slowly spread around the system via trader networks in such a way that the system incrementally self-organises into a bubble-build-up phase. Peaks triggered by endogenous shocks also have slower relaxation rates as the system self-organises to unwind back down toward the Triple Point. In contrast, exogenous shocks, that is, shocks triggered by a single overwhelming external event, have a more rapid build-up to the Critical Point as well as a faster relaxation rate afterwards.

In discussing the search for a narrative to explain what he calls the *Great Recession*, Yergin (2009) argues that the normal oscillation between 'fear and greed' characterising stock markets following EMH at the Triple Point was shifted in favour of greed by the historically low cost of risk. Add to this, the US Federal Reserve's Chairman Ben Bernanke's comment in his 23 October 2009 speech to the US Board of Governors of the US Federal Reserve System:

> The extraordinary pressure on financial firms last fall underscored how profoundly interconnected firms and markets are in our complex, global financial system.

The 'fear, greed, risk' dynamics of single human traders now interacted among thousands of traders across a vast strongly-interconnected global Financial Services system. Bernanke ended his speech by calling on the US Congress to legislate supervision capable of regulating vast complex interconnected systems rather than just individual CEOs, boards, or banks. While individual banks failed, it was global complex-system failure that caused the Great Recession, not individual agents – whether persons or banks. At the Triple Point isolated, rational, individual behaviour rules the day. But as a market progresses toward the Critical Point, connectivity and scalability dynamics increasingly dominate to produce the extreme outcome at the Critical Point – a historically significant market crash.

Referring to our earlier use of Bak's sandpile avalanches in Chapter 5, the *tension* basis of sand cascades from small to large is obvious in stock trading – *'fear and greed'*. The *connectivity* of sand grains is what the US Federal Reserve Bank's Chairman Bernanke points to: *profound interconnectivity of firms throughout the global Financial Services system*. We do not deny that resilience engineering in many ways has to bear down on individual CEOs, Boards, and Banks. But resilience engineering is meaningless if it doesn't produce global-system resilience. Bernanke stresses both individual and system in his speech. In the present chapter we attempt to focus on both individual entities of the system as well as the overall system.

Creating Resilience

We agree: the magic of free markets as Yergin (2009) calls it, should not be eradicated, as some national governments seem to prefer. At the Triple Point, and for some distance out towards the Critical Point, the free market mantra should dominate. But as greed and connectivity begin to dominate market behaviour – *that is, once the R1 tipping point is reached* (see Chapter 7 for details) – resilience interventions need to be triggered to counteract the impact of Financial Engineering to stabilise the system until the global system transitions back below the $R1$ tipping point, that is, to a more safe and healthy system state. In short, financial markets should be free to *resiliate* freely and without the impact of lobbying.

As mentioned in the beginning of this chapter, we advocate that resilience interventions should be (mostly) automatically activated once the tipping point is crossed. Needless to say, activation may either be fully automatic, or depend on timely human decisions by regulatory and/or political entities. Thus, while we agree with Lord Turner (who was the UK's chief financial regulator at the beginning of 2010) that special regulatory powers by what he terms a macro-prudential committee are needed to 'prevent asset price bubbles' (Giles, 2010) we also see the need to *change* his call for it to meet 'twice a year … with power to pull a macro-prudential lever … if appropriate'. As stated, we favour *automatic* imposition of the levers with power given to the committee to negate them some time later if other conditions warrant such modifying actions. Our principle is that *formulaic automaticity comes first*, with judgements imposed by (political) regulators coming later if further negative financial circumstances justify.

But is *global* system resilience engineering impossible? We wonder. Most nations do not like the idea of being regulated by other nations; witness the difficulties in dealing with global warming and nuclear weapons. Nobel Laureate James Tobin introduced his *Tobin tax* on global currency exchange in 1978, but it has never taken hold even though there have been many strong advocacies in favour of it.

Also, note that while we use the term *formulaic automaticity* just above, we do not go so far as to set forth specific formulas – this needs to become the focus of technically trained financial resilience engineers. Most of the formulas that we propose should be activated at the $R1$ tipping point, are often not much more that simple ratios between increasing risk and increasing upfront cost to Financial Services firms in the form of taxes on bonuses, capital reserve requirements, diminished risk-taking, and so forth.

We categorise our resilience interventions into the following types: First, *Information Disclosure, Against Individual Agents, Systemic Interconnectivities,* and *Other.* Analogous to people driving cars, we divide *Individual Agents* into *Reckless Endangerment* (reckless Financial Engineering given easily available knowledge) and *Insurance* (protection against unexpected anomalies given increased risk, leverage, use of mortgage-backed securities, securitisation packages, and so on). Generally speaking, our proposed resilience interventions fall into two broad categories:

a. Information Disclosures and Sharing. The goal is to reduce noise in favour of improved, more trusted information, so that traders will change behaviour, leading the overall system to transition to the EMH at the Triple Point.
b. Impose Restrictions on Trading Activities and/or Trigger Insurance Frameworks. These should be designed to dampen negative impact on the overall system. These interventions are called for once the system crosses the $R1$ tipping point (as indicated by the Hurst exponent).

Creating Resilience via Information Disclosure

In this Section we build on the idea that noise in balanced EMH trading is really random information. But if new or relevant information from trusted information sources is expensive, it becomes difficult for it to readily play its role as noise in EMH trading. As a result, we suggest the following list of information disclosures as interventions that are likely to facilitate the emergence of resilience:

1. General: Any kind of increased *information disclosure (from trusted information sources)* about herding and bubble-build-up indications will increase the likelihood that traders will retreat from rule-based herd trading back to Triple Point trading. For example, a news report that traders appear to be copying a questionable trading rule rather than independently searching for better information about fundamentals (that is, about the fundamental value of a firm's stock).

2. Any kind of information that helps uncover the $R1$ tipping point: for example, the *Financial Times* (*FT*) could offer a weekly or daily report about Hurst exponents for various stock markets, industries, and other categories worldwide.

3. Information that would disrupt/reduce *trader-homogeneity and, thus, bolster efficient-market trading*. For example, once the Hurst exponent suggests that $R1$ has been passed, the *FT* could pay more attention to reports suggesting traders are putting more emphasis on rule-copying rather than relevant information about fundamentals.

4. Information that helps regulatory entities define conditions and *causes* of reckless endangerment. For example, banks keep offering subprime teaser loans even though they see that interest rates are rising; Goldman Sachs' covering up the size of the Greek sovereign debt could have been reported earlier; the amount of loans based on mortgage-backed securities rather than solid assets could have been systematically reported some four years before the 2007 credit crisis began.

5. Disclose information about estimated losses rather than wait until losses have occurred. This has been proposed by The International Accounting Standards Board (Sanderson and Hughes, 2009).

6. Information relevant to mitigating *endogenous* shocks:
   - Calls for information about events similar to the tiny initiating events and spirals we highlighted in the 35 years leading up to the 2007 liquidity crisis.
   - Makes easier comparisons of future events to those we have highlighted from the past 35 years.

7. Disclosures related to questionable management, Board Director interlocks, M&A mistakes, Ponzi-use of bonuses to inappropriately motivate unacceptable levels of *risk taking* (cf. Bernanke, 2009).

8. Information about horizontal connectivity among firms and banks and how this is tracked and controlled across countries, as opposed to just watching and regulating vertically within individual firms (cf. Bernanke, 2009).

9. Information suggesting that the end of the herding-based bubble-build-up (as indicated by the Hurst exponent) may be in sight. This possibility is suggested by Grech and Pamula (2008), although Fagura et al. (2010) suggest that more research is needed.

Creating Resilience against Individual Agents Such as Traders and Banks

In this Section we suggest a number of resilience interventions designed to be applied to individuals and banks. These interventions should only be instigated once trading volatilities have increased past the R1 tipping point.

Reckless Endangerment   It was common knowledge in the US that the US Federal Reserve Bank's Discount rate was 1.25 per cent in December 2001 and 0.75 per cent in November 2002 and that it would have to be raised to protect against inflation. Teaser loans lasted five years, and people could buy homes without declaring income or credit score. Therefore we define leveraging mortgage-backed securities at 10, 20, 30, 40, and even up to 50 to 1 (in Europe) as demonstrable reckless endangerment. For the most part, the US Federal Reserve Bank or European Central Bank can define recklessness well in advance of negative outcomes, since certain types of investment behaviours are well known to be associated with high risk. Given this, we suggest that the following interventions should be made once the R1 tipping point has been identified:

1. All further risk-taking should automatically be gradually reduced by *formulaic resilience interventions* against Financial Engineering activities as the bubble progresses, until risk is fully mitigated by valid risk-free assets and/or cash-based escrow accounts.
2. All bonuses, salaries, and other bank equities should be put in escrow (for example, in a government-held account) until appropriate revenues have been achieved; The US Federal Reserve Bank's chairman Bernanke stated that 'compensation practices ... have led to misaligned incentives and excessive risk-taking ... The Federal Reserve is working to ensure that compensation packages appropriately tie rewards to long-term performance and do not create undue risk ...' (Labaton, 2009).

3. Warnings by the relevant regulatory entity (or entities) should be given as early as possible so that all bank parties may start taking appropriate action. The importance of early warning systems is described in Chapter 7 of this book.

4. Executives, financial engineers and traders directly responsible for recklessness should be held accountable for the impact of their reckless behaviour. For example, they should lose their positions and be held liable dating back to the start of reckless judgements; direct superiors up to and including CEOs should also be fired (also see Record, 2010).

5. Reckless endangerment becomes more relevant when banks participate in an *'asset-price bubble'* of the kind that Mishkin (2009: 11) describes as '... a credit boom bubble ...' based on wilful bank-caused (endogenous) high-leveraged lending practices relying on questionable assets like mortgage-backed securities, as opposed to a *'pure irrational exuberance bubble'* (like the one based on dot.com stocks in the late 1990s) which was exogenous. By the US Securities and Exchange Commission's account, this includes 'securitised – or repackaged – pools of subprime mortgages' and re-securitised collateralised debt obligations (CDOs) backed by mortgages and other loans held by a bank. By 2007 ~74 per cent of CDOs had failed and ~86 per cent of bonds backed by mortgage-backed securities had failed (Chung, 2010).

Insurance   Like any other insurance, an insurance to improve resilience is defined by both the total financial liability and some measure of the likelihood of crash-causing anomalies. As asset liability and risk of anomalies increase, insurance costs increase.

1. Designated high-risk banks and other players (such as US firms, AIG, Fanny Mae, Freddie Mac) should pay into an *insurance* fund; payments should be made *before* bonus payments; payments should be progressive (that is, the percentage increases as risk increases) and begin at the $R1$ tipping point. After some time, say six months (our latest research suggests that $R1$ comes about one year before the primary crash (Fagura et al., 2010), payments should amount to some percentage of the prior year's bonus pool, say 90 per cent (the latter used to be the highest income tax in the US).

2. At the G20 Finance Ministers' meeting at St Andrew's, Scotland, the former UK Prime Minister Gordon Brown and the former Chancellor of the Exchequer Alistair Darling once again brought up the so-called *Tobin tax*, also referred to as the 'casino tax' (cf. Aitken, 2009; Eaglesham et al., 2009). This tax would be applied to all financial transactions. Our view is that this tax should be introduced as a resilience intervention at the $R1$ tipping point and then increased from miniscule to substantial. But the US, Canada, and IMF immediately dismissed this proposal. The current proposal is a compromise: the tax would not be applied until well after the $R1$ tipping point – too late to offer any useful advantage.

3. Raise *borrowing costs* for high-risk taking banks. This accounts to Yergin's easy credit which is at the base of Shilleresque bubbles in real estate, energy, and so on. McCormack (2009: 13) suggests that the best solution would be 'a tax on short-term debt, especially short-term debt of financial institutions'.

4. What *The Economist* (2009b) calls capital ratios (also-called living wills or catastrophe insurance) should be imposed in a timely fashion to penalise risky behaviours.

5. Mervyn King, Governor of the Bank of England, suggested that when the State acts as lender of last resort it would be appropriate to 'impose a windfall tax on bank profits or a *Tobin tax* on transactions' (*The Economist*, 2009a: 68).

6. US President Obama has proposed a bank fee based on the amount of a financial firm's liabilities.

7. The US regulatory entity, FDIC, takes the position of wanting a depositors' insurance fee levied against bank compensation plans that promulgate high-risk strategies while paying out bonuses (similar to Ponzi schemes) well before knowing whether the risk-taking produced profits or losses (Chung and Guerrera, 2010).

8. Forced reduction in *leverage*. This accounts to Yergin's 'too much leverage' and would be defined in terms of a ratio to a bank's secure assets, or increased insurance against the leverage.

9. Limits on *short selling* (there are several kinds), or increased insurance to cover downside risks and costs; thus at $R1$:

- 'Naked' short selling (no assets involved) is stopped.
- Computerised short-selling is stopped.
- The next worst type of short-selling, and so on, are stopped as the bubble grows.
- Though not necessarily short-selling, we also call for growing limits on high-speed autonomous computerised trading after $R1$ since the tremendous volume – largely unsupervised – can have significant market impacts; this surely gains importance after the flash crash of 7 May 2010 in which the DJI lost 998 points in less than half an hour.

10. *Shadow operations* by individual hedge funds should be disclosed at the tipping point. Insurance costs should then progressively be applied.
11. More attention focused on giving shareholders more relevant information and more *responsibility/authority/control* over CEO Ponzi bonus-giving coupled with inordinate risk-taking behaviours. For example, Ken Lewis – then CEO of Bank of America (a large US Financial Services firm) – reduced shareholder value by buying Merrill Lynch after the latter had paid out billions of US dollars in bonuses to traders who engineered incredible failures After $R1$, shareholders should have required voting rights on increasing risk, leverage, and so on.

Systemic Interconnectivities All resilience interventions aimed at systemic interconnectivities should be applied after the $R1$ tipping point. Some our proposed interventions include:

1. Growing limits on *securitisation repackaging.* The repackaging process is systemic since one financial firm will sell to other firms on a global basis. Growing limits *on non-cash securitisation repackaging;* also systemic, as described above.
2. Tracking the *contagion of FE formulas.* When their contagion level reaches a specific level of trader homogeneity, insurance costs against systemic collapse need to be imposed, as called for by US Federal Reserve Bank's Chairman (Bernanke, 2009).
3. Tracking growth and operations of *shadow trading* (which includes hedge and private equity funds owned by deposit

banks, as well as computer stock trading, derivative markets, and unreported, unregulated 'private' stock markets). Interventions must be made in these markets at the tipping point.

4. Make markets more susceptible to transient *international* changes. As sovereign debt defaults by governments (for example, Greece, Dubai, Argentina, Iceland, Ireland and so on), bank failures and so forth – become more likely, traders and firms at risk must be required to build up insurance, reduce leverage, and restrict short selling.

Other Possible Interventions We agree with Rostowski (2010; Finance Minister of Poland) that when interest rates are extremely low – as they were in 2001–2003 when the bubble started (and as they are now in 2010) – then they should be automatically raised once $R1$ is reached, and be raised even higher if the rates have already been raised off their lows. The so-called 'Greenspan put', which kept interest rates very low longer than necessary, is to be avoided like the plague.

Going back to our description of the liquidity crisis example in Chapter 7, resilience engineering interventions in both the Reckless Endangerment and Insurance categories could have been made anywhere along the progression towards the full blown crisis. In other words, our intent is to make it obvious in how many ways and stages scalability dynamics became operative well before the onset of the 2007 crisis. Resilience interventions could be imposed before $R1$, but this would interfere with unimpaired operation of efficient-market trading. To avoid this, we think it best not to impose resilience interventions until after the $R1$ tipping point. By this standard, virtually all of the currently suggested anti-crisis measures are too strong before $R1$ and too weak afterwards.

Returning to the Normal

All the foregoing *formulaic resilience interventions* should be scaled back when the volatility PL disappears (that is, when the Hurst exponent goes below ½) as the system reverts back below the $R1$ tipping point. In our view, the interventions could zero out well before the Triple Point is reached.

While we believe our instigation of *formulaic resilience engineering interventions* after the R1 tipping point is better than those measures currently being discussed (many of which we have already alluded to in our earlier discussion of specifics), there are still two problems to worry about:

1. How to prevent financial engineers from 'playing' the system (like people constantly finding loopholes in tax regulations)?
2. How many formulas should be initiated at the R1 tipping point and how should they be designed?

First, playing the system: even Christopher Cox, the US Securities and Exchange Commission's Chairman, admitted that regulation methods were fundamentally flawed from the beginning (Labaton, 2008; see also Kauffman, 2008). Over the past couple of years the US SEC has been pummelled for its failure to take early note of (1) the US Madoff Ponzi scheme, and (2) its inability to hire staff competent to understand, let alone regulate, Financial Engineering. The key question pertaining to the latter is why talented financial engineers who can make millions of US dollars on Wall Street should be willing to settle for making thousands of US dollars working for the Securities and Exchange Commission? And if this is indeed the case, how can the SEC or the US Federal Reserve Bank be able to outwit the gaming of the financial engineers on Wall Street or in hedge funds scattered around the world?

The second problem is the formulas. In order to counteract the forces of bubbles the design of the formulas should as far as possible be secret and randomly changeable/applicable so that even if gaming takes place it is more like roulette than playing against a relatively unchanging and very public tax code. Although we do not try to design these formulas in the present chapter, we do suggest an initial focus on simple ratios connected to numbers based on the size of matters easily identified after the R1 tipping point is passed including:

1. A progressive tax; that is, an increasing percentage of bonuses and bank equities are put in temporary escrow. Time in escrow increases in direct ratio with (risk leverage) of all loans based on non-secure assets.

2. The more that bank decisions pertain to higher (non-secure asset-based risk) leverage, the more intervention measures need to focus on executive management and Board of Directors' decisions.
3. Banks defined as high-risk based on (non-secure asset-based risk) leverage should pay into an insurance fund before any major bonus or dividend payouts.
4. A progressive Tobin-like tax is imposed on all financial transactions – admittedly, an intervention that may be difficult to execute. Banks defined as high-risk would face progressively increasing taxes on the amount borrowed as (non-secure asset-based risk) leverage increases.
5. As any kind of leverage is increased, a progressive bank fee is imposed.
6. As the ratio of (non-secure asset-based risk) leverage increases relative to secure assets, banks progressively pay increasing amounts into an insurance fund.
7. As the ratio of (non-secure asset-based risk) leverage increases relative to secure assets, bank insurance fees are progressively increased. Short selling, shadow operations, and non-human computerised trading are progressively limited.
8. All of the above are increased as their ratio to non-secure-asset based loan securitisations increases.
9. As combinations of very low interest rates coupled with teaser loans appear, loans based on non-secure assets are reduced.
10. Interest rates on loans based on any kind of asset are automatically progressively raised as volatilities increase after $R1$.

Summary

In this chapter we have proposed a set of formulaic resilience engineering interventions to be applied at the $R1$ tipping point to offset irrational exuberance by financial engineers and all their formulas. We offer a wide variety of ~40 different resilience interventions. We begin by suggesting 10 bearing on basic measures for improving *Information Disclosure* interventions –

the better information quality and relevance is, the more relevant and timely are the resilience interventions. As our review of the build-up to the 2007 liquidity crisis in Chapter 7 illustrates, relevant information is available, as long as one only knows what to look for. Then we point out that, like driving a car, there are obvious cases of deliberate *Reckless Endangerment* – five resilience interventions here – as well as 11 interventions to build up *Insurance* against unpredictable anomalies. Then, because modern financial markets are so *Systemically Interconnected* around the globe, we offer five resilience interventions that become relevant as interconnectivity builds. Finally, and most importantly, we believe that the only viable solution, given the computer-based speed at which Financial Engineering quant formulas operate, is to set up resilience formulas that automatically take hold at $R1$ – only *resilience engineering formulas* can beat *financial engineering* formulas.

There are now many books and even more articles and newspaper stories calling for heightened regulation. Some of these are summarised in *The Economist*'s report on financial risk (2010). However, there is no evidence that any currently proposed solutions do what our approach does: leave the banking system as currently regulated – that is, relatively unregulated – as long as it stays at the Triple Point, which is to say, as long as it stays with balanced greed, risk, and uncertainty and behaves according to Bachelier's random walk and Fama's EMH. Our approach is more like progressive income tax: poor people pay no tax, but as people earn more income they pay progressively more tax. Our resilience interventions only apply after the $R1$ tipping point is reached in stock trading volatilities. They only start being imposed after EMH market behaviour ceases to dominate and traders start to lose their heterogeneity. In other words, herding behaviour starts, buy–sell rules become more similar, bank and global connectivities multiply (for example, the US firm Lehman Brothers was comprised of nearly 3,000 legal entities in dozens of countries when it went bankrupt, *The Economist*, 2010: 15), *and* if the Hurst exponent indicates that a bubble-build-up toward a crash has begun.

All of the 12 examples we mentioned earlier are easily reduced to formulas involving progressive bank-cost increases as risk increases, that is, they are based on ratios involving readily identified risks taken against readily identifiably asset quality. We recognise that there always is the danger and temptation of Financial Services firms attempting to hide risky actions. We argue, however, that *hidden risk can be measured as the ratio of incentives offered relative to the total known assets of a Financial Services firm*. Either this is a valid approach or incentives are paid out without merit (risky or not). Or, alternatively, incentives are based on hidden assets and shadow banking.

The proposed resilience interventions have to be imposed as early as possible as markets transition from random-walk EMH (Bachelier, 1914; Fama, 1970) trader behaviour at the Triple Point, to bubble-build-ups based on trader herding behaviour leading up to the Critical Point (that is, stock market crash; Yalamova and McKelvey, 2011). However, other interventions should not be made too early. The key question is: how early we can get resilience dynamics introduced to offset the various dynamics underlying the build-up toward the Critical Point? A recent study by Wasden (2010) finds that two-thirds of the cues pointing to the failure of Bear Stearns and Lehman Brothers came after the fact of their failure As a consequence, no interventions imposed by regulators or politicians had any chance of being preventive. As discussed in Chapter 7, our proposed $R1$ *tipping-point indicator* should help to detect bubble-build-ups in a proactive manner. Even if only *half* of the 10+ per cent market crashes are predicted early on, the reduction of the impact on the broader economies would translate into savings of billions of US dollars.

One final note: If our proposed resilience interventions are put into practice after the $R1$ tipping point, the ongoing discussions about whether to reinstate some form of the Glass-Steagall Act become moot and *too-big to fail* also becomes moot. Deposit-taking banks would not need to be separated from high-risk investment banks (the Volcker Rule), or what former US Treasury Secretary Nicholas Brady calls the *shadow banking system* (2010). The *five mini-Goldmans* proposed by MIT professor, Simon Johnson (Task, 2010) would not be needed.

Chapter 10

The Ability to Regulate, Govern and Control Financial Services Systems

Gunilla Sundström and Erik Hollnagel

The description of the financial crisis in Section I showed how macro-prudential regulatory functions failed to detect and act on early warning signals. These early warning signs included the bankruptcy filing by US based subprime lender New Century Financial Corporation in April 2007 and the drastic change in borrowers' default rates in the US. Micro-prudential regulators, such as UK's Financial Services Authority, also failed to recognise that the risk exposure of the individual firm corresponded to the risks experienced by the majority of firms of the Financial Services system. One reaction to this inability to detect early warning signs has been the call to establish regulatory bodies focused on monitoring and identifying systemic risk. In Europe, the so-called de Larosière report proposed a European Systemic Risk Council with a charter to 'form judgments and make recommendations on macro-prudential policy, issue risk warnings, compare observations on macro-economic and prudential developments and give direction on these issues' (de Larosière, 2009: 44). The US Treasury Department made a similar recommendation in March 2009, by calling for the establishment of a Systemic Risk Regulator. In July 2010, the proposed Financial Reform was signed into US legislation. Similarly, the UK has reformed its banking system with the banking act issued in 2009, and continues to introduce other measures to make it easier to understand the impact of banking failures on individual customers. Finally, in

September 2010, the Basel III accord was introduced by the Group of Governors and Heads of Supervision of the International Bank for Settlements (www.bis.org). This accord was accepted by the G20 countries in October 2010 and foresees that banks must hold capital equal to at least 7 per cent of its assets.

In order for any suggested regulatory change to have practical consequences, the regulatory entities in charge of the monitoring and identification of risk must have a clear view of the following:

- The Financial Services system to be monitored, and in particular an understanding of how the system functions when it is in a good (that is, healthy) state.
- The Key Performance Indicators (KPI) that make it possible to monitor the system's state. The KPIs must include early warning indicators that reliably show when a system is in danger of transitioning into an unhealthy state. This could be a state where a governmental bailout and or an acquisition by another entity is the only remedy.
- The viable actions needed to govern and control the system to either maintain a healthy state and/or to guide the system from an unhealthy to a healthy state.

In Section I, we described how Financial Services systems can be viewed as open dynamic systems and how negative (dampening) and positive (amplifying) feedback loops can be used to understand the behaviour of these systems including their interactions with the environments. In this chapter we leverage some key learnings from cybernetics (for example, Ashby, 1956), the study of feedback systems (for example, Åström and Murray, 2009) and resilience engineering (Hollnagel, Woods and Leveson, 2006) in a discussion of regulation, governance and control of Financial Services systems.

Regulation, Governance and Control

The three terms regulation, governance and control are related. In the Financial Services industry, *regulation* is defined as the policies established to ensure that risks associated with Financial Services systems are identified and properly managed. One of the key regulatory requirements in the Financial Services industry is the capital requirement for banks as defined by the new Basel III accord (Wellink, 2010). As previously mentioned, the Basel III accord, requires banks to keep a reserve capital equal at least 7 per cent of their assets. The actually required capital reflects a particular bank's risk exposure. If the capital reserve decreases but the risk exposure remains the same, the bank becomes undercapitalised and the risk of failure is assumed to increase. As its predecessor, that is, Basel II, the Basel III accord is focused on individual components of the Financial Services system and not on the overall system (see Aglietta and Scialom, 2009 for comments on Basel II).

The objective of *governance* is to establish policies, processes and procedures facilitating the achievement of various goals, such as regulatory goals, corporate goals, or goals related to specific functions and even individual programmes and projects. Finally, *control* denotes the activities needed to keep a system's performance within a defined range to make sure it can achieve its objectives.

Corporate management comprises the set of control activities used in Financial Services system to keep a corporation on track to achieve its business objectives. Financial Service system typically have the following management, regulatory and control functions:

- The individual firm's management function that is responsible for achieving business results while managing risks at the same time. Ultimately, the Chief Executive Officer is accountable for a particular firm's risk exposure and for the strategies used to manage risks.
- At least one micro-prudential regulatory function responsible for the oversight of individual firms. In the US, the Office of Comptroller of the Currency is an example of a micro-prudential regulatory body.
- A macro-prudential regulatory function responsible for the oversight of the overall Financial Services system. Central Banks, such as the Bank of England and the European Central Bank, are examples of macro-prudential regulators. The Financial Stability Oversight Council (FSOC), introduced

as part of the 2010 US financial reforms, is responsible for monitoring of systemic risk. The council members include all major US regulatory bodies and it is charged with monitoring of both banking and non-banking entities (for example, insurance entities) that compose the US Financial Services system.

Successful corporate management requires that acceptable ('healthy') and unacceptable ('unhealthy') system states can be identified. Figure 10.1 illustrates three states of a Financial Services system (cf. Sundström and Hollnagel, 2006) called the healthy, the unhealthy, and the catastrophic states. While a system may make a transition between healthy and unhealthy states in both directions, a transition into a catastrophic state is irreversible and fatal. (In practice, a system that has failed catastrophically can sometimes be acquired by another firm, or bailed out by a national government.) Figure 10.1 also illustrates how state variables (that is, revenue, leverage and capital reserves) can be used to characterise and possibly identify the state of the Financial Services system.

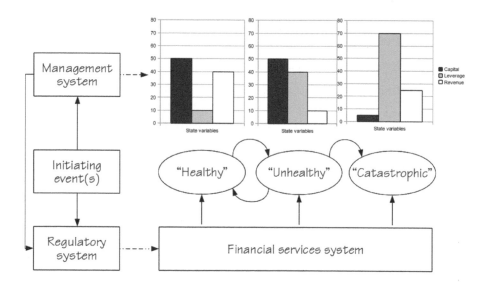

Figure 10.1   **Three generic system states identified by state variables**

The concept of state variables is used in control theory to denote variables that can be used to predict and anticipate a system's future states (cf. Åström and Murray, 2009). In order for the three state variables shown in Figure 10.1 to be used for a shared view of how the Financial Services system can be monitored, they must first be developed into a set of Key Performance Indicators that support a state identification. Figure 10.1 also suggests that external events can impact the Financial Services system directly and force the regulatory system to a reactive response. Ideally, such events should be anticipated by the management system, thus enabling the regulatory system to be proactive and prepare the Financial Services system for the impact of the negative events.

About Regulation

In the classical text on cybernetics, Ashby (1956) defined the problem of regulation as the design of a regulatory mechanism that is coupled to the environment in such a way that the variability of state variables falls within the range defined as a healthy state. A regulating mechanism can accomplish this in two basic ways:

- The first is by 'physically' blocking or prevention of the impact of events. A Financial Service system's management team could for instance react to market changes by deciding to divest a certain type of business. If a particular type of business no longer is part of the system, then any events in the firm's environment related to the business have been 'physically' blocked. However, it does not necessarily mean that these events cannot have an impact on other parts of the firm's business.
- The second is by anticipating the events and proactively prepare for them. For example, a Financial Service system's management team may believe that credit card delinquency rates will increase due to deteriorating general economic conditions. To prepare for this impact of the economic downturn, the management team can decide to increase their loss reserves and to proactively shed risky customers.

In order to be in control of something it must be possible to influence system behaviour in such a way that performance stays within acceptable limits, in other words that performance meets expectations. How such expectations are defined depends on the type of system that is being monitored and controlled. Systems engineering, leveraging control theory (for example, Åström and Murray, 2009), has focused on developing sensor technology to support monitoring and control of complex systems. Examples are found in telecommunications, power generation and aviation, among others. However, the technology used to control system behaviour is often embedded in components, and therefore invisible to the human eye – and sometimes also incomprehensible to the human mind. An airplane cockpit is an environment with embedded technology that helps human pilots to control how the airplane flies. Another example is control rooms as they can be found in power plants, air traffic control systems, and railway dispatch centres.

In Financial Services systems such sophisticated real-time monitoring and control systems are typically found layered around key technology components, such as the networks leveraged by the global Financial Services system. The trading floors in various financial markets are probably the most sophisticated real-time systems of any industry, at least in terms of the amount of data acquired and processing speeds. However, the primary purpose of these systems is to support a key function, that is, trading, and not to monitor and or control any particular Financial Services system. In fact, financial markets and processes can range from being primarily paper based and manual to be sophisticated and highly automated. Traditional retail deposit banking in small banks represent one end of the spectrum, whereas global equity and forex markets represent the other end. The fact that *centralised* real-time monitoring and control environments, such as financial cockpits or control rooms, have not been developed for management in the same way they have been developed for trading, poses a serious challenge for those that are chartered with management and control of Financial Services systems, in particular those entities chartered with monitoring systemic risk.

A Framework for Applying Resilience Engineering to Financial Services Systems

An organisation, or, a system is said to be resilient if it is able to adjust its functioning prior to, during, or following changes and disturbances, so that it can sustain required operations under both expected and unexpected conditions (Hollnagel, 2011a). Resilience engineering argues that this requires four abilities, namely the ability to respond, the ability to monitor, the ability to learn, and the ability to anticipate. There are significant similarities between the four abilities and the description of ambidexterity in Chapter 8, cf. also Grote (2009).

During the financial crisis, Financial Service systems failed on at least three of the four abilities. They did not anticipate the crisis, and did not monitor it well enough; indeed, many financial institutions completely failed to detect it, until it was too obvious to overlook. Neither did they respond very well, and the inability to find effective responses is something that even today affects how quickly a new sustainable condition can be established. It can be argued that all this partly was due to a failure to learn the proper lessons from the past. Hopefully, the learning from the recent crisis will be better.

In the present discussion, we shall focus on the abilities to monitor and to respond. But it must be pointed out that a more thorough analysis will show that none of the four abilities can be seen independently of any of the others. The reasons for focusing on monitoring and response are: (1) that an effective response is necessary in order to maintain control of a situation, and (2) that effective monitoring, and detection of potentially critical developments, is a precondition for effective responses. Monitoring allows responses to be more timely, or even to be proactive, hence prevents the system from falling behind the development of events and slide into a state where the ability to function deteriorates because of never-ending fire-fighting.

Monitoring – Knowing What to Look For

The key requirements for a resilient system are that it knows what to look for (that is, which state variables to monitor) and that is able to use that information to mitigate the impact of both external and internal events – at least in real time and preferably proactively, before something happens. Knowing what to look for is essential for every organisation regardless of whether it is described as being resilient or as a High Reliability Organisation. But knowing what to look for (and actually looking for it) is not enough. Once something has been found, actions or responses must follow. The quality of the responses can be characterised in terms of the level of control they represent that is, the orderliness of the responses. Levels of control can range from a scrambled (panic like) response, over an opportunistic response, such as the herding behaviour described in Chapter 7, to tactical and strategic responses. (A more detailed description of levels of control can be found in, for example, Hollnagel and Woods, 2005.)

The importance of monitoring means that effective data collection and data analysis become two key attributes of Financial Services systems practices. The data collection and data analysis capabilities must focus on two different sources of data and information:

- First, data from the system's operating environment. For a Financial Services system this environment could be defined by geographical borders, by business interests, by a risk profile, or by a decision to monitor all events belonging to a certain category such as actions by central banks.
- Second, data from the Financial Services system itself. This is important, both because extreme events may occur within the system itself (think of rogue traders), but also because there nearly always are unrecognised, tight couplings between internal and external developments.

Since it is impossible to look at all events and all data, there must be a way to select or filter from the inputs that which should be analysed further (sorting the wheat from the chaff, so to speak). The selection or filtering must clearly be systematic and orderly, which means that it must be based on a set of articulated assumptions or hypotheses. This is usually referred to as a model of the Financial Services system, that is, a simplified representation of the system, its environment and its key attributes. The key feature of any model is that it helps to determine what data are relevant and how the data should be organised, processed, and interpreted. The models therefore also affect how the system boundaries are defined and guide the way assessments and decisions are made. Regulators of the Financial Services industry should be very careful in determining what kind of model to use, from both a macro- and a micro prudential perspective. Otherwise they may miss important signals, and therefore fail to respond at the right time. Responses that rely on an incorrect model are likely to be ineffective and may even be directly wrong or harmful.

An example of a commonly used model is the Efficient Market Hypothesis (EMH), explained in Chapter 7. In order to determine how close the market is to the Triple Point, it is necessary to evaluate the risk-to-fundamentals ratio and the noise-to-information trading ratio. The EMH also makes clear how these data are to be processed and interpreted. Another example is the discussion in Chapter 6, where sandpiles and Dragon Kings were used to explain how Financial Service systems work. These, and many other, analogies are clearly very helpful to describe what happens, and not least to communicate the description to others. They can also be helpful in suggesting 'signatures' for which one should look out, as in the case of the EMH. The analogies are useful and important because they are empirically salient, but properly speaking they do not constitute models of the underlying phenomenon or of how Financial Services systems work. They are therefore not well suited to suggest different or novel indicators, or to suggest more detailed and more predictive indicators.

The Institutional and the Functional Perspectives

As discussed in Section I, the two different ways to look at the Financial Services industry are the institutional and the functional perspectives (cf. Merton, 1995). An institutional perspective focuses on the institution providing Financial Services, whereas a functional perspective focuses on functions performed in the Financial Services industry. Merton (1995) argued that a functional perspective was better suited to support a global perspective. The primary argument was that the rapid technological changes and the increased integration of financial markets were stronger determinants of performance than the way a specific institution chose to implement – or institutionalise – specific functions. Functional perspectives, or functional models, have also been shown to be very useful in other complex industries such as telecommunications, power generation and aviation (for example, Hollnagel et al., 2008). This is because functional models facilitate views across idiosyncrasies, such as implementation details associated with individual system components. A functional perspective can therefore enable an understanding of how such systems work by identifying risks across Financial Services institutions. It can also help individual companies to identify the critical dependencies between their business model and the behaviour of key functions of the overall Financial Services system (cf. Sundström and Hollnagel, 2011).

A central question is, of course, which functions should be monitored and how it can be done. One possibility is to consider the functions suggested by Merton (1995) and Merton and Bodie (2005):

- Clearing and Settlement – payment services related to the exchange of goods and services.
- Risk Management – manage uncertainty, develop and implement mitigation strategies and action plans.
- Transfer of Economic resources – facilitating the flow of economic resources resulting in efficient use of capital.
- Resource Pooling – bringing together economic resources from various sources to create large capital pools.
- Information Sharing – provide stakeholders with access to risk ratings, price information and other types of information perceived as critical for making decisions in (or about) Financial Services systems.
- A mechanism to deal with asymmetric availability of information, that is, one party has access to certain information whereas other parties do not.

Economic systems are typically described using the concepts of demand and supply, and the few existing definitions of Financial Services systems illustrate that. For example, Schmidt and Tyrell (2004: 21) defined a Financial Services system 'as the interaction between supply of and the demand of the provision of capital and other finance related resources'. It therefore makes sense to include Demand and Supply functions in a proposed Financial Services systems model as well.

A Function-Based Representation

Sundström and Hollnagel (2011) suggested the Functional Resonance Analysis Method (FRAM; Hollnagel, 2004) as a modelling and analysis framework for the Financial Services industry. The FRAM provides a way to identify how functions are coupled and how such couplings may lead to unexpectedly large performance variability and unexpected outcomes. These unexpected outcomes can be either positive (that is, desirable) or they can be negative (that is, undesirable). Without going into details, the FRAM comprises the following steps:

- Identify the functions that are necessary and sufficient for the everyday performance of the system or organisation; characterise each by six basic aspects (Inputs, Outputs, Preconditions, Resources, Time, and Control).
- Describe the actual/potential variability of 'foreground' and 'background' functions (process and context). Consider both everyday and excessive variability.
- Define functional resonance based on potential/actual dependencies (couplings) among functions.
- Propose ways to monitor and dampen performance variability (indicators, barriers, design/modification, and so on), cf. Chapter 9.

Example of a Functional Model

In the notation shown in Figure 10.2, each function (shown as a hexagon) is described using two or more of the following attributes:

- Input (I): The signal or change that activates the function and/or the entity or substance that is used or transformed to produce the output. The input constitutes the link to upstream functions. In Financial Services the input can be a request for transfer of economic resources.

- Output (O): That which is the result of the function, for example, an entity or a state change. Constitutes the links to downstream functions. A change in credit rating is an example of an output.
- Preconditions (P): System conditions that must be fulfilled before a function can be executed. A risk assessment is an example of a precondition.
- Resources (R): That which is needed or consumed by the function when it is active (for example, matter, energy, competence, software, capital, manpower). Funds and capital are key resources of any Financial Services system. Resources that are not actually depleted by the function but which are required for the execution nonetheless (for example, competence), are called execution conditions.
- Time (T): Temporal aspects that affect how the function is carried out (constraint, resource).
- Control (C): That which supervises or regulates the function. This can be plans, procedures, guidelines or other functions. A firm's management and/or regulatory entities are examples of controls.

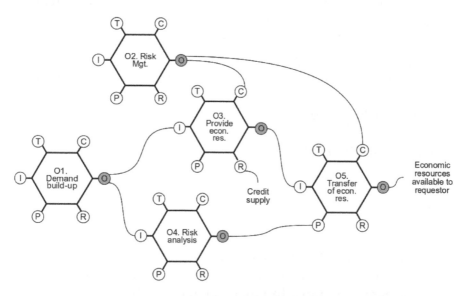

Figure 10.2   **An example of a FRAM-based functional representation**

Figure 10.2 illustrates possible interdependencies between a number of economic functions – Demand (O1) and Supply (O3), and Global Credit Market system functions – Risk Management (O2), Risk Analysis (O4) and Transfer of Economic Resources (O5).

The lines represent typical couplings between the different functions. For example, a demand can initiate the provision of economic resources (O1:O – O3:I) as well as a risk analysis (of the requestor) (O1:O – O4:I). The availability of the necessary resources can be a 'signal' to initiate the transfer (O3:O – O5:I). But the transfer is also contingent upon a satisfactory result of the risk analysis (O4:O – O5:P). Finally, both the provision of economic resources and the transfer are guided or controlled by risk management (O2:O – O3:C) and (O2:O – O5:C).

The couplings or interdependencies represented by the lines show how one function may affect another, not in terms of causes and effects but in terms of propagation of variability. If, for instance, the risk management function (O2) does not work satisfactorily (for example, that either anticipation or monitoring do not work well), then the provision (O3) and the transfer (O5) of economic resources may both be adversely affected. Figure 10.2 illustrates how it is possible to elaborate on the principles of Figure 3.4. Instead of describing dependencies among functions simply in terms of inputs and outputs, the FRAM allows a greater level of precision by referring to the six aspects (I, O, P, R, T, and C).

The brief description of how the FRAM can be used to identify functional interdependencies illustrates how the Financial Services system can be modelled. As argued above, it is important for any regulatory body to articulate their view of the Financial Services system, that is, their model. The model is important both for proposing specific KPIs, for instance related to the variability of essential functions, and also for deciding on specific types of intervention, such as damping the variability of a function. Without such an explicit approach for how to look at Financial Services systems, discussions about increased regulatory scrutiny will remain rather meaningless.

Responding – Knowing What to Do

The model of the Financial Services system is necessary for monitoring, since it makes clear *what* one should look for. If the dynamics of the model are reliable, then it may also help to decide *when* to look, that is, how frequently performance indicators should be sampled. In the same way, the model is necessary for responding, since it can be used to develop responses that are effective, that is, that lead to the desired changes in the state of the system.

This can be done because the model shows both *which* functions are coupled, and *how* they are coupled. A coupling between the output of one function and the precondition of another is, for instance, different from a coupling between the output of one function and the control or resource of another. Since the couplings typically will be 1:n or n:1 – or even n:n – rather than 1:1, the model may also show how variability may build up and propagate through the system and thereby lead to precarious instabilities.

Resilience engineering also makes clear that responding is not just knowing what to do, but also being able to do it. This includes issues such as *when* to respond (the timing of an economic stimulus, for instance), as well as when to cease or terminate a response. Other issues have to do with the availability of *resources*. Several scholars pointed out that regulators were not able to perform effectively because they lacked sufficient resources. For example, whereas almost everyone on Wall Street had access to a Bloomberg Terminal, the same was not the case at the Securities and Exchange Commission. Yet another concern is how to maintain the capability to respond, and how to verify that this is the case.

Conclusions

Like any other complex system, effective governance and control of Financial Services system requires that regulatory bodies establish a view of what the Financial Services system is. This view will help identify the key behaviours that must be monitored, leveraging various types of performance and early warning indicators. Leveraging learnings from resilience engineering, systems engineering, and control theory, we suggested that a functional perspective of the Financial Services system would be most useful. The primary reason being that a functional perspective refers to functions rather than structures; it is therefore not constrained by the geographical and institutional boundaries that normally are taken for granted. To be effective on a global basis, regulatory entities should ideally have a high-level shared functional view of the global financial system. We suggest that the FRAM approach provides such a framework for the development of a functional shared model. As illustrated above, simple models provide a view of key couplings and interdependencies and thus help to understand how variability in or changes to a function might propagate through the system as a whole. A functional model might in other words help to anticipate impact of changes in functions on the overall system.

Resilience engineering proposes that a resilient system must have four capabilities in order to remain in a 'healthy' state over time. These are the abilities to respond, to monitor, to learn, and to anticipate. Anticipation is not possible without a view of what might happen, that is, some view of the future. This in turn must make use of experience from what has happened in the past. Responding is likewise not possible without predicting what the consequences of a response may be, nor without monitoring how the current situation develops. Learning is required for many reasons, one being that learning reduces the probability of repeating actions which either are ineffective or which have a negative impact on the overall system.

There is not yet an agreed view of which performance indicators ought to be shared by various regulatory entities, beyond capital requirements and leverage ratio described in the Basel III accord. However, as illustrated in Figure 10.1, it is crucial to have a view of what performance indicators must be monitored to determine the current state of a Financial Services system. Without such a view, the regulatory system itself is not resilient and thus prone to ad hoc changes in reaction to various events! In addition to monitoring such indicators as capital ratios and leverage, the governance and control of Financial Services systems could do worse than adopt the suggestion that resilient systems have the four mentioned capabilities. As a result, identifying ways to monitor the presence of these capabilities in Financial Services systems, for instance by using the technique of the Resilience Analysis Grid (Hollnagel, 2011b), would complement the more traditional ways to monitor the behaviour of Financial Services systems. A key goal of resilience engineering is to contribute to the development of such techniques by leveraging the experiences and applications from other domains.

Epilogue: Financial Markets and the Law of Requisite Variety

Erik Hollnagel and Gunilla Sundström

Economics has a history that is replete with examples of system descriptions and, in modern times, elaborate hypotheses about how financial markets work. The range of hypotheses and models goes from Adam Smith's (and Kenneth Arrow's) 'invisible hand', over Bachelier's 'random walk' and Mandelbrot's fractals, to Fama's 'Efficient Market Hypothesis' and the contemporary system models à la econophysics. But as recent developments have shown, they all have their shortcomings.

In the current situation it has gradually, but grudgingly, been acknowledged that simple models do not work. Simple models, such as the Domino model (Heinrich, 1959; but a first mentioning from 1934) are attractive because they are simple, and therefore easy to understand and communicate. But even if simple models are adorned with mathematics, they remain fundamentally simple. And it is very hard, if not impossible, to understand something if the explanation is based on a single – and simple – principle. The Law of Requisite Variety, briefly mentioned in Chapter 3, tells us that in order to decrease the variety of the outcomes (of a system) we need to increase the variety of the controller of that system (Ashby, 1956). A more down-to-earth formulation is that a good regulator of a system must be a model of the system (Conant and Ashby, 1970), which means that the regulator or controller of a system must possess a complexity that is comparable to that of the system. Or even more simply: it is impossible to control a system that one does not understand. The reason for this is straightforward: if a system can respond in ways that the controller cannot represent or imagine, then the system becomes impossible to control and manage.

When the variety of the regulator cannot be increased, the Law of Requisite Variety actually suggests another solution, namely to reduce the variety of the process. One way of doing that is to impose more regulation, or more constraints, on the markets in the hope that this will make them more predictable. (The reaction to the 2010 flash-crash was the same: trading curbs or 'circuit breakers'.) But this type of solution just exacerbates the problem, since we can only effectively regulate or constrain a process if we know how it works. Which brings us back to square one.

A major theme in this book has been the *What-You-Look-For-Is-What-You-Find* principle. This argues that the assumptions that people, or communities, make when they try to understand events to a large extent determine the outcome, that is, the 'meaning' of the events, and therefore also the actions taken in consequence of that. The responses made to the recent global financial crisis clearly demonstrated that different schools of thinking understood the crisis in different ways, and that there was little agreement between them. On the one side was the Keynesian supply philosophy, and on the other the philosophy of fiscal prudence. However, in both cases the proposed solution or 'control' was a single response type – spend or save – which cannot reasonably be said to match the complexity of the financial systems – even if we do not understand them fully. The fact that there can be two opposite responses to the very same situation corroborates the *What-You-Look-For-Is-What-You-Find* principle, and that different underlying models lead to different diagnoses of the same manifestations.

From Structure to Function

Models, in economy, finance and elsewhere, typically describe a system's components and how they are organised, that is, the structure. In the spirit of system thinking, some models go on to consider how the components function, including how such functions may fail. For example, the failure of the regulator, the failure of the credit rating service, the failure of the 'rogue' trader.

Resilience engineering, however, takes a slightly different stance. It does not try to model the system, that is, the economy, by presuming that it must have a certain architecture or certain mathematical characteristics (fractal, power functions, self-organisation, or whatever the flavour of the day is). Resilience engineering rather looks at the functions or abilities that are necessary for any system to be in control of what it does, no matter what that is. It turns out that it is sufficient to have four different abilities, namely the abilities to respond, to monitor, to learn, and to anticipate.

In addition to the four abilities it is also necessary to understand the process dynamics, which in this case means having a model or description of how the financial system works. This model is necessary to design and prepare responses (since their efficacy can be tested on the model), to know what to look for in terms of monitoring (for example, the bubble phenomenon), to learn what works and what does not, and to anticipate – which essentially is making predictions or projections. Learning is also necessary to update and improve of the model, as well as to make improvements on the operational level, such as why a specific response was effective. (Note, however, that if we only know that a response was effective but not why, learning may be incorrect since it may reflect spurious correlations.)

The intention of this book, as described in the Prologue, has been to gain a better understanding of the recent crisis of the global financial system and try to answer the questions about what actually happened, why it happened, and whether it could happen again. The two first questions have been answered in various chapters, but not with a simple and single answer, as there probably is none. The third question is easier to answer. Yes, it could happen again – and given that we only have a partial understanding of how the global financial system works, it is likely to happen sooner rather than later.

This, of course, raises the question of what we should do to prevent the occurrence of a similar, or even more serious, crisis. Resilience engineering cannot give a ready-made answer to that, simply because resilience engineering is a science of control and management rather than a theory of financial markets. (On the other hand, there probably are no ready-made answers anyway.)

resilience engineering can, however, help us to understand better how we can regulate the financial markets, and also tell us what qualities a model of the market must have. We do need to bootstrap a model, but it should reflect the actual phenomena rather than dogma. We know that effects are non-linear and that outcomes are emergent rather than resultant. We know that the Financial Services system is tightly coupled and partly intractable. We know that things go right and things go wrong in the same basic manner. And we know that effective control requires a judicious mixture of feedforward and feedback. With this knowledge in hand we at least know how to proceed, which must be better than the current opportunistic search for the financial market version of the philosopher's stone.

References

Acharya, V.V. and Richardson, M. (2009). Causes of the financial crisis. In V.V. Acharya and M. Richardson, *Restoring Financial Stability: How to Repair a Failed System*. New York, USA: John Wiley and Sons. [1]

Adamski, A.J. and Westrum, R. (2003). Requisite imagination: the fine art of anticipating what might go wrong. In E. Hollnagel (ed.), *Handbook of Cognitive Task Design* Mahwah. NJ: Lawrence Erlbaum Associates, pp. 193–220. [3]

Adrian, T. and Shin, H.S. (2007). Liquidity and Leverage. Unpublished paper, Federal Reserve Bank of New York and Princeton University. (September). [6]

Aglietta, M. and Scialom, L. (2009). *A Systemic Approach to Financial Regulation – A European Perspective*. Foundation for European Progressive Studies, June. [10]

Aitken, M. (2009). Chancellor's Plan for Tax on Banks Gets Shot Down at G20. *Daily Record* (8 November), http://www.dailyrecord.co.uk/news/uk-world-news/2009/11/08/chancellor-s-plan-for-tax-on-banks-gets-shot-down-at-g20–86908–21806360/ (accessed 9 November 2009). [9]

Akerlof, G.A. (1970). The market for 'lemons': quality uncertainty and the market mechanism. *The Quarterly Journal of Economics*, 84, 488–500. [3]

Alvarez-Ramirez, J., Alvarez, J., Rodriguez, E. and Fernandez-Anaya, G. (2008). Time-varying Hurst Exponent for US stock markets. *Physica A*, 387, 6159–6169. [4] [7]

Andergassen, R., Reggiani, A. and Nijkamp, P. (2004). Firm Dynamics and Self-Organised Criticality, Entrepreneurship and Regional Economic Development: A Spatial Perspective, 292–313. Available at SSRN: http://ssrn.com/abstract=1496751. [6]

Andriani, P. and McKelvey, B. (2007). Beyond Gaussian averages: extending organization science to extreme events and power laws. *Journal of International Business Studies*, 38, 1212–1230. [4] [5] [7]

Andriani, P. and McKelvey, B. (2009). From Gaussian to Paretian thinking: causes and implications of power laws in organizations. *Organization Science*, 20, 1053–1071. [4] [5] [7]

Andriani, P. and McKelvey, B. (2011). From skew distributions to power-law science. In P. Allen, S. Maguire and B. McKelvey (eds), *Handbook of Complexity and Management*. London: Sage. [4]

Argote, L. (1982). Input uncertainty and organizational coordination in hospital emergency units. *Administrative Science Quarterly*, 27, 420–434. [8]

Argote, L., Fichman, M. and Turner, M.E. (1989). To centralize or not to centralize – the effects of uncertainty and threat on group-structure and performance. *Organizational Behavior and Human Decision Processes*, 43, 58–74. [8]

Ashby, W.R. (1956). *An Introduction to Cybernetics*. London, UK: Chapman and Hall. [3] [10]

Åström, K.J. and Murray, R.M. (2008). *Feedback Systems: An Introduction for Scientists and Engineers*. Princeton, NJ: Princeton University Press. [3][10]

Auerbach, F. (1913). Das Gesetz Der Bevolkerungskoncentration. *Petermanns Geographische Mitteilungen*, 59, 74–76. [4]

Axtell, R.L. (2001). Zipf distribution of US firm sizes. *Science*, 293, 1818–1820. [4]

Bachelier, L. (1914). *Le Jeu, la Chance et la Hasard*. Paris: E. Flammarion. [4] [5] [9]

Bak, P. (1996). *How Nature Works: The Science of Self-organized Criticality*. New York: Copernicus. [4] [5]

Bak, P., Tang, C. and Wiesenfeld, K. (1987). Self-organized criticality: an explanation of 1/f noise. *Physical Review Letters*, 59, 381–384. [4] [5]

Baker, D. (2009). *Plunder and Blunder: The Rise and Fall of the Bubble Economy*. Sausalito, CA: PoliPoint Press. [5]

Banerjee, A.V. (1992). A simple model of herd behavior. The *Quarterly Journal of Economics*, 107, 797–817. [7]

Barabási, A.-L. (2002). *Linked: The New Science of Networks*. Cambridge, MA: Perseus. [4] [5]

Basili, M. and Zappia, C. (2003). Probabilistic versus Non-probabilistic Decision Making: Savage, Shackle and Beyond. University of Siena Economics Working Paper No. 403. [7]

Baskin, K. (2005). Complexity, stories and knowing. *Emergence: Complexity and Organization*, 7, 32–40. [5]

Bastiaensen, K., Cauwels, P., Sornette, D., Woodard, R. and Zhou, W.-X. (2009). The Chinese equity bubble: ready to burst. Available at http://arxiv.org/abs/0907.1827 and http://ssrn.com/abstract=1479479. [6]

Baum, J.A.C. and McKelvey, B. (2006). Analysis of extremes in management studies. In D.J. Ketchen, Jr. and D.D. Bergh (eds), *Research Methodology in Strategy and Management*, vol. 3, 123–196. Elsevier Ltd. [7]

Becker, M.C. (2004). Organizational routines: a review of the literature. *Industrial and Corporate Change*, 13, 643–677. [8]

Benner, M.J. and Tushman, M.L. (2003). Exploitation, exploration, and process management: the productivity dilemma revisited. *Academy of Management Review*, 28, 238–256. [8]

Bernanke, B.S. (2009). Financial Regulation and Supervision after the Crisis: The Role of the Federal Reserve. *Financial Times* (23 October). [9] http://www.ft.com/cms/s/0/1f0e0df8-bfd2-11de-aed2-00144feab49a.html (accessed 24 October 2009). [9]

Beunza, D. and Stark, D. (2004). Tools of the trade: the socio-technology of arbitrage in a Wall Street trading room. *Industrial and Corporate Change*, 13, 369–400. [8]

Beunza, D. and Stark, D. (2005). How to recognize opportunities: heterarchical search in a trading room. In K. Knorr Cetina and A. Preda (eds), *The Sociology of Financial Markets*. Oxford: Oxford University Press, pp. 84–101. [8]

Bikhchandani, S., Hirshleifer, D. and Welch, I. (1992). A theory of fads, fashion, custom, and cultural change as informational cascades. *Journal of Political Economy* 100, 992–1026. [7]

Black, F. (1972). Capital market equilibrium with restricted borrowing. *Journal of Business*, 45, 444–455. [4]

Black, F. and Scholes, M. (1973). The pricing of options and corporate liabilities. *Journal of Political Economy*, 81, 637–654. [4] [5]

Blanchard, O. (1979). Speculative bubbles, crashes and rational expectations. *Economic Letters*, 3, 387–389. [7]

Blanchard, O. (2008). The financial crisis: initial conditions, basic mechanisms, and appropriate policies. Munich lecture. (November). [6]

Blanchard, O. and Watson, M. (1982). Bubbles, rational expectations and financial markets. In P. Wachtel (ed.), *Crises in the Economic and Financial Structure*, 295–316. Lexington, MA: D.C. Heathland Co. [7]

Bollerslev, T. (1986). Generalized autoregressive conditional heteroscedasticity. *Journal of Econometrics*, 31, 307–327. [7]

Brady, N.F. (2010). Refocus the Regulatory Debate on Essentials. *Financial Times* (4 January), http://www.ft.com/cms/s/0/3e8f914e-f969-11de-8085-00144feab49a.html (accessed 15 February 2010). [9]

Brock, W.A. (2000). Some Santa Fe scenery. In D. Colander (ed.), *The Complexity Vision and the Teaching of Economics*. Cheltenham, UK: Edward Elgar, pp. 29–49. [4]

Broekstra, G., Sornette, D., Zhou, W.-X. (2005). Bubble, critical zone and the crash of Royal Ahold. *Physica A: Statistical Mechanics and its Applications*, 346(3–4), 529–560. [6]

Brown, S.L., Eisenhardt, K.M. (1995). Product development – past research, present findings, and future directions. *Academy of Management Review*, 20, 343–378. [8]

Brügger, U. (2000). Speculating: work in financial markets. In H. Kalthoff, R. Rottenbrug and H.-J. Wagener (eds), Facts and figures: economic representations and practices, *Ökonomie und Gesellschaft, Jahrbuch 16.* Marburg: Metropolis, pp. 229–255. [8]

Brunnermeier, M.B. (2009). Deciphering the liquidity and credit crunch 2007–2008. *Journal of Economic Perspectives*, 23, 77–100. [3]

Brunnermeier, M.K. (2001). *Asset Pricing under Asymmetric Information: Bubbles, Crashes, Technical Analysis, and Herding*. Oxford: Oxford University Press. [4] [5] [7] [9]

Buckley, W. (1968). Society as a complex adaptive systems. In Buckley, W. (ed.), *Modern Systems Research for the Behavioral scientist*. Chicago: Aldine Publishing Company, p. 490–513. [2]

Burns, T. and Stalker, G.M. (1961). *The Management of Innovation*. London: Tavistock Publications. [8]

Cajueiro, D.O. and Tabak, B.M. (2004). The Hurst Exponent over time: testing the assertion that emerging markets are becoming more efficient. *Physica A*, 336, 521–537. [7]

Cajueiro, D.O., Tabak, B.M. and Werneck, F.K. (2009). Can we predict crashes? The case of the Brazilian stock market, *Physica A*, 388, 1603–1609. [9]

Canning, D., Amaral, L.A.N., Lee, Y., Meyer, M. and Stanley, H.E. (1998). A power law for scaling the volatility of gdp growth rates with country size. *Economics Letters*, 60, 335–341. [4]

Cavagna, A. (1999). Irrelevance of memory in the minority game. *Phys. Rev. E59*, R3783-R3786. [6]

Challet, D. and Zhang, Y.-C. (1997). Emergence of cooperation and organization in an evolutionary game. *Physica A: Statistical and Theoretical Physics*, 246 (3–4), 407–418. [6]

Challet, D., Marsili, M. and Zhang, Y.-C. (2005). *Minority Games: Interacting Agents in Financial Markets*. USA: Oxford University Press. [6]

Cherns, A. (1987). Principles of sociotechnical design revisited. *Human Relations*, 40, 153–162. [8]

Chung, J. (2010). Regulator has Sights Trained on Bank Sales Practices. *Financial Times* (15 January), http://www.ft.com/cms/s/0/8a3910f8–0174–11df-8c54–00144feabdc0.html (accessed 15 February 2010). [9]

Chung, J. and Guerrera, F. (2010). FDIC Eyes Linking Levies to Bank Pay. *Financial Times* (7 January), http://www.ft.com/cms/s/0/1b733a94-fb26–11de-94d8–00144feab49a.html (accessed 15 February 2010). [9]

Clegg, S.R., Courpasson, D. and Phillips, N. (2006). *Power and Organizations. Foundations for Organizational Science*. London: Sage. [8]

Conant, R.C. and Ashby, W.R. (1970). Every good regulator of a system must be a model of that system. *International Journal of Systems Science*, 1 (2), 89–97. [3]

Cooper, G. (2008). *The Origin of Financial Crises*. New York: Vintage Books. [4][5]

Czarnecki, L. Grech, D. and Pamula, G. (2008). Comparison study of global and local approaches describing critical phenomena on the Polish stock exchange market. *Physica A*, 387, 6801–6811. [4]

de Larosière, J. (2009). The High-Level Group of Financial Supervision in the EU. Brussels, 25 February 2009. [3][10]

Di Matteo, T., Aste, T. and Dacorogna, M.M. (2005). Long-term memories of developed and emerging markets: using the scaling analysis to characterize their stage of development. *Journal of Banking and Finance*, 29, 827–851. [4] [7]

Dudley, W.C. (2010). Asset bubbles and the implications for central bank policy. Speech by Mr William C. Dudley, President and Chief Executive Officer of the Federal Reserve Bank of New York, at the Economic Club of New York, New York City, 7 April. [6]

Eaglesham, J., Giles, C. and Hollinger, P. (2009). Brown Retreats from Tobin Tax Proposal after US-led Backlash. *Financial Times* (9 November), 1. [9]

Ee, K.H. and Xiong, K.R. (2008). Asia: a perspective on the sub-prime crisis. *Finance and Development* (June), 19–23. [10]

Eisenhardt, K.M. and Martin, J.A. (2000). Dynamic capabilities: what are they? *Strategic Management Journal*, 2, 1105–1121. [8]

Emery, F. and Trist, E. (1965). The causal texture of organizational environments. *Human Relations*, 18, 21–32. [2]

Engle, R.F. (1982). Autoregressive conditional heteroscedasticity with estimates of variance of United Kingdom inflation. *Econometrica*, 50, 987–1008. [7]

Engle, R.F. and Bollerslev, T. (1986). Modeling the persistence of conditional variances. *Econometric Reviews*, 5, 1–50. [7]

Eom, C., Choi, S., Oh, G. and Jung, W.-S. (2008). Hurst Exponent and prediction based on weak-form efficient market hypothesis of stock markets. *Physica A*, 387, 4630–4636. [7]

Fagura A., Kunal, K. and Poag, Z. (2010). Can the Hurst Exponent Be Used As A Predictive Measure of Asset Bubble Formations and Eventual Bursts? Research paper, UCLA Anderson School of Management, Los Angeles, CA. [7] [9]

Fama, E.F. (1965). Random walks in stock prices. *Financial Analysts Journal*, 21(5), 55–59. [4] [7]

Fama, E.F. (1970). *Efficient Capital Markets: A Review of Theory and Empirical Work. Journal of Finance*, 21, 383–417. [Pt I] [4] [5] [6] [7] [9]

Fama, E.F. (1998). Market efficiency, long-term returns, and behavioral finance, *Journal of Financial Economics*, 49, 283–306. [Pt I] [4] [7]

Federal Reserve Bank of San Francisco. (2005). Economic Letter. Available at: http://www.frbsf.org/publications/economics/letter/2005/el2005–17.html. [5]

Feldman, S.P. (2004). The culture of objectivity: quantification, uncertainty, and the evaluation of risk at NASA. *Human Relations*, 57, 691–718. [8]

Ferraro, F., Pfeffer, J., Sutton, R.I. (2005). Economics language and assumptions: how theories can become self-fulfilling. *Academic Management Review*, 30, 8–24. [8]

Financial Stability Board (2009). www.financialstabilityboard.org. [3]

Forrester, J.W. (1968). Market growth as influenced by capital investment. *Industrial Management Review*, 9 (2). [3]

Foster, J.B. and Magdoff, F. (2009). *The Great Financial Crisis*. New York: Monthly Review Press. [5]

Fox, J. (2009).*The Myth of the Rational Market*. New York: Harper Business. [4] [5]

Gabaix, X. (1999). Zipf's Law for cities: an explanation. *Quarterly Journal of Economics*, 114 (3), 739–767. [6]

Gallati, R.R. (2003). *Risk Management and Capital Adequacy*. McGraw-Hill, New York. [8]

Gell-Mann, M. (1988). The concept of the institute. In D. Pines (ed.), *Emerging Synthesis in Science*, pp. 1–15. Boston, MA: Addison-Wesley. [4]

Gell-Mann, M. (2002). What is complexity? In A.Q. Curzio and M. Fortis (eds), *Complexity and Industrial Clusters*, pp. 13–24. Heidelberg, Germany: Physica-Verlag. [4]

Geller, R.J., Jackson, D.D., Kagan, Y.Y. and Mulargia, F. (1987). Earthquakes cannot be predicted. *Science*, 275 (5306), 1616. [6]

Ghashghaie, S., Breymann, W., Peinke, J., Talkner, P. and Dodge, Y. (1996). Turbulent cascade in foreign exchange markets. *Nature*, 381, 767–770. [4]

Ghysels, E., Santa-Clara, P. and Valkanov, R. (2005). There is a risk-return trade-off after all. *Journal of Financial Economics*, 76, 509–548. [7]

Giles, C. (2010). Turner Calls for Powers to Deflate Asset Bubbles. *Financial Times* (27 January), http://www.ft.com/cms/s/0/d2bf7a9c-0b4c-11df-9109-00144feabdc0.html (accessed 15 February 2010). [9]

Glaser, P. (2009). Fitness and Inequality in an Increasing-returns World. Term paper, Complexity Leadership course, Spring Quarter, UCLA Anderson School of Management, Los Angeles, CA. [4]

Granger, C.W.J. and Joyeux, R. (1980). An introduction to long-memory time series models and fractional differencing. *Journal of Time Series Analysis* 1, 15–30. [7]

Grech, D. and Mazur, Z. (2004). Can one make a crash prediction in finance using the local Hurst Exponent idea? *Physica A*, 336, 133–145. [7]

Grech, D. and Pamula, G. (2008). The Local Hurst Exponent of the financial time series in the vicinity of crashes on the Polish stock exchange market. *Physica A*, 387, 4299–4308. [7] [9]

Greenspan, A. (1996). The Challenge of Central Banking in a Democratic Society. Remarks at the Annual Dinner and Francis Boyer Lecture of the American Enterprise Institute for Public Policy Research, Washington DC, 5 December. [4]

Grossman, S. and Stiglitz, J.E. (1980). On the impossibility of informationally efficient markets. *American Economic Review*, 70, 393–408. [7]

Grote, G. (2009). *Management of Uncertainty – Theory and Application in the Design of Systems and Organizations*. London: Springer. [8]

Grote, G., Weichbrodt, J.C., Günte,r H., Zala-Mezö, E. and Künzle, B. (2009). Coordination in high-risk organizations: the need for flexible routines. *Cognition Technology and Work*, 11, 17–27. [8]

Gupta, A.K., Smith, K.G. and Shalley, C.E. (2006). The interplay between exploration and exploitation. *Academic Management Journal*, 49, 693–706. [8]

Hale, A.R. and Swuste, P. (1998). Safety rules: procedural freedom or action constraint? *Safety Science*, 29, 163–177. [8]

Halpin-Healy, T. and Zhang, Y.-C. (1995). Kinetic roughening, stochastic growth, directed polymers and all that, *Physics Reports*, 254, 215–415. [6]

Heinrich, H. (1959). *Industrial Accident Prevention: A Scientific Approach*. 4th Edition. NY: McGraw-Hill. [1] [3]

Herald Tribune. (2009). What is $100 million? (6 October). [8]

Hickson, D.J., Hinings, C.R., Lee, C.A., Schneck, R.E. and Pennings, J.M. (1971). Strategic contingencies theory of intraorganizational power. *Administratice Science Quarterly*, 16, 216–229. [8]

Hirshleifer, D. and Teoh, S.H. (2003). Herd behaviour and cascading in capital markets: a review and synthesis. *European Financial Management*, 9, 25–66. [4] [5] [7] [9]

Holland, J.H. (1995). *Hidden Order: How Adaptation Builds Complexity*. Reading, MA: Addison-Wesley. [5] [9]

Holland, J.H. (2002). Complex adaptive systems and spontaneous emergence. In A.Q. Curzio and M. Fortis (eds), *Complexity and Industrial Clusters*. Heidelberg, Germany: Physica-Verlag, pp. 25–34. [5] [9]

Hollnagel, E. (2004). *Barriers and Accident Prevention*. Aldershot, UK: Ashgate. [10]

Hollnagel, E. (2009a). *The ETTO Principle: Efficiency Thoroughness Trade-Off. Why Things that Go Right Sometimes Go Wrong*. Farnham, UK: Ashgate. [Pt I] [2] [3]

Hollnagel, E. (2009b). The four cornerstones of resilience engineering. In C.P. Nemeth, E. Hollnagel and S. Dekker (eds), *Preparation and Restoration*. Farnham, UK: Ashgate (p. 117–134). [Pt I]

Hollnagel, E. (2011a). Prologue: the scope of resilience engineering. In E. Hollnagel, J. Pariès, D.D. Woods and J. Wreathall (eds), *Resilience Engineering in Practice: A Guidebook*. Farnham, UK: Ashgate. [10]

Hollnagel, E. (2011b). Epilogue: RAG – the resilience analysis grid. In E. Hollnagel, J. Pariès, D.D. Woods and J. Wreathall (eds), *Resilience engineering in practice: A guidebook*. Farnham, UK: Ashgate. [10]

Hollnagel, E. and Speziali, J. (2008). *Study on Developments in Accident Investigation Methods: A Survey of the 'State-of-the-Art'*. SKI Report 2008, 50. [1] [3]

Hollnagel, E. and Woods, D.D. (2005). *Joint Cognitive Systems: Foundations of Cognitive Systems Engineering*. Boca Raton, FL: CRC Press/Taylor and Francis Group. [10]

Hollnagel, E., Woods, D.D. and Leveson, N. (2006). *Resilience Engineering: Concepts and Precepts*. Ashgate, Aldershot UK. [2] [3] [8][10]

Hollnagel, E., Pariès, J., Woods, D.D. and Wreathall, J. (eds). (2011). *Resilience Engineering in Practice*. Farnham, UK: Ashgate. [3]

Hollnagel, E., Pruchnicki, S., Woltjer, R. and Etcher, S. (2008). A functional resonance accident analysis of Comair flight 5191. Paper presented at the *8th International Symposium of the Australian Aviation Psychology Association*, Sydney, Australia. [10]

Hurst, H.E. (1951). Long-term storage capacity of reservoirs. *American Society of Civil Engineers Transactions*, 116, 770–808.

Ishikawa, A. (2006). Pareto Index induced from the scale of companies. *Physica A*, 363, 367–376. [4]

Jaffe, D.M. (2008). The US Subprime Mortgage Crisis: Issues Raised and Lessons learned. Prepared for the Commission on Growth and Development and the World Bank. 11 April, Workshop on Fiscal and Monetary Policies and Growth. [1]

Jensen, H.J. (1998). *Self-Organized Criticality*. Cambridge: Cambridge University Press. [6]

Jiang, Z.-Q. and Zhou, W,-X. (2007). Scale invariant distribution and multifractality of volatility multipliers in stock markets. *Physica A*, 381, 343–350. [4]

Jiang, Z.-Q., Zhou, W.-X., Sornette, D., Woodard, R., Bastiaensen, K. and Cauwels, P. (2010). Bubble diagnosis and prediction of the 2005–2007 and 2008–2009 Chinese stock market bubbles. *Journal of Economic Behavior and Organization*. [4] [6]

Jogi, P. and Sornette, D. (1998). Self-organized critical random directed polymers. *Phys. Rev.*, E57, 6931–6943. [6]

Johansen, A. and Sornette, D. (2001). Large stock market price drawdowns are outliers. *Journal of Risk*, 4 (2), 69–110. [6]

Johansen, A. and Sornette, D. (2006). Shocks, crashes and bubbles in financial markets. *Brussels Economic Review* (Cahiers economiques de Bruxelles), 49 (3/4). Available at: http://papers.ssrn.com/paper.taf?abstract_id=344980. [6]

Johansen, A. and Sornette, D. (1998). Stock market crashes are outliers. *European Physical Journal B*, 1, 141–143. [7]

Joint Center for Housing Studies, Harvard University (2008). [1]

Kardar, M. and Zhang, Y.-C. (1987). Scaling of directed polymers in random media, *Physics Review Letter*, 58, 2087–2090 (1987).

Kauffman, S.A. (1993). *The Origins of Order*. New York: Oxford University Press. [5]

Kaufmann, D. (2008). Capture and the Financial Crisis: An Elephant Forcing a Rethink of Corruption? Blogs.worldbank.org, http://blogs.worldbank.org/governance/capture-and-the-financial-crisis-an-elephant-forcing-a-rethink-of-corruption (accessed 15 February 2010). [9]

Keene, S. (No date). Political Economy: Evolutionary Economics, Power Laws and Evolutionary Modeling. Lecture slides; School of Economics and Finance, University of Western Sydney. [7]

Kennedy, P. (2003). *A Guide to Econometrics*. 5th Edition. Cambridge, MA: MIT Press. [4]

Kindleberger, C.P. (2000). *Manias, Panics, and Crashes: A History of Financial Crises*. 4th Edition. New York: John Wiley. [10]

Klayman, J. and Ha, Y.-W. (1987). Confirmation, disconfirmation and information in hypothesis testing. *Psychological Review*, 94, 2, pp. 211–228. [2] [3]

Knight, F.H. (1921). *Risk, Uncertainty, and Profit*. Boston: Houghton Mifflin. [7]

Knorr Cetina, K. and Bruegge,r U. (2002). Global microstructures: the virtual societies of financial markets. *American Journal of Sociology*, 107, 905–950. [8]

Krugman, P. (2009). *The Return of Depression Economics and the Crisis of 2008*. New York: Norton and Co. [5]

Labaton, S. (2008). S.E.C. concedes oversight flaws fueled collapse. *New York Times* (26 September), http://www.nytimes.com/2008/09/27/business/27sec.html (accessed 15 February 2010). [9]

Labaton, S. (2009). Fed plans to vet banker pay to discourage risky practices. *New York Times*, (23 October), http://www.nytimes.com/2009/10/23/business/23pay.html?_r=1 (accessed 15 February 2010). [9]

Lahart, J. (2007). In time of tumult, obscure economist gains currency. *Wall Street Journal*, (18 August), http://online.wsj.com/article/SB118736585456901047.html. [5]

Laherrère, J. and Sornette, D. (1999). Stretched exponential distributions in nature and in economy: fat tails with characteristic scales. *European Physical Journal*, B2, 525–539. [6]

Leamer, E.E. (1990). In D.F. Hendry, E.E. Leamer, and D.J. Poirier (eds), A conversation on econometric methodology. *Econometric Theory*, 6, 171–261. [4]

LeBaron, B. (2001). Volatility Magnification and Persistence in an Agent Based Financial Market. Working paper, Brandeis University. [7] [9]

Levy, M. and Solomon, S. (1997). New evidence for the power-law distribution of wealth. *Physica A*, 242, 90–94. [4]

Lintner, J. (1965). The valuation of risk assets and the selection of risky investments in stock portfolios and capital budgets. *Review of Economics and Statistics*, 47, 13–37. [4]

Lorenz, E.N. (1972). Predictability: Does the Flap of a Butterfly's Wings in Brazil Set Off a Tornado in Texas? Paper presented at the 1972 meeting of the American Association for the Advancement of Science. Washington, DC. [5]

Luhmann, N. (1979). *Trust and Power*. Chichester: Wiley. [8]

MacKenzie, D.A. (2006). *An Engine, not a Camera: How Financial Models Shape Markets*. Cambridge, MA: MIT Press. [6] [8]

MacKenzie, D. and Millo, Y. (2003). Constructing a market, performing theory: the historical sociology of a financial derivatives exchange. *American Journal of Sociology*, 109 (1): 107–145. [10]

Malevergne, Y., Pisarenko, V.F. and Sornette, D. (2005). Empirical distributions of log returns: between the stretched-exponential and the power law? *Quantitative Finance*, 5 (4), 379–401. [6]

Malevergne, Y., Saichev, A. and Sornette, D. (2008). Zipf's Law for Firms: Relevance of Birth and Death Processes. Available at: http://ssrn.com/abstract=108396. [6]

Mandelbrot, B.B. (1963a). The variation of certain speculative prices. *Journal of Business*, 36, 394–419. [4] [5]

Mandelbrot, B.B. (1963b). New methods in statistical economics. *Journal of Political Economy*. 71, 421–440. [4]

Mandelbrot, B.B. (1982). *The Fractal Geometry of Nature*. New York: Freeman. [4] [7]

Mandelbrot, B.B. (1997). *Fractals and Scaling in Finance, Discontinuity, Concentration, Risk*. New York: Springer-Verlag. [5] [7]

Mandelbrot, B.B. and Hudson, R.L. (2004). *The (mis)Behavior of Markets: A Fractal View of Risk, Ruin, and Reward*. New York: Basic Books. [4] [7]

Mantegna, R.N. (1991). Lévy walks and enhanced diffusion in Milan stock exchange. *Physica A*, 179, 232–242. [4]

Mantegna, R.N. and Stanley, H.E. (2000). *An Introduction to Econophysics: Correlations and Complexity in Finance*. Cambridge, UK: Cambridge University Press. [4]

March, J.G. (1991). Exploration and exploitation in organizational learning. *Organizational Science*, 2, 71–87. [8]

Marris, P. (1996). *The Politics of Uncertainty: Attachment in Private and Public Life*. London: Routledge. [8]

Maruyama, M. (1963). The second cybernetics. *American Scientist*, 51, 164–179. [3] [5]

Maskawa, J. (2007). Stock price fluctuations and the mimetic behaviors of traders. *Physica A*, 382: 172–178. [7]

McCormack, R. (2009). A tax on short-term debt would stabilise the system. *Financial Times*, (16 December), 13, http://www.ft.com/cms/s/0/0bf7a7d8-ea7a-11de-a9f5-00144feab49a.html (accessed 15 January 2010). [9]

McKelvey, B. (2011). Fixing the UK's economy. In J. McGlade, M. Strathern and K. Richardson (eds), *Complexity in Human and Natural Systems*. Litchfield Part, AZ: ISCE Publishing. [4]

McKelvey, B., Lichtenstein, B.B. and Andriani, P. (2011). When systems and ecosystems collide: is there a law of requisite fractality imposing on firms? In M.J. Lopez Moreno (ed.), *Chaos and Complexity in Organizations and Society*. Madrid, Spain: UNESA. [4]

Merton, R.C. (1995). A functional perspective of financial intermediation. *Financial Management*, 24 (2), p. 21–41. [2][6][10]

Merton, R.C. and Bodie, Z. (1995). A conceptual framework for analysing the financial environment. In D.B. Crane, K.A. Froot, S., P. Mason, A.F. Perold and R.C. Merton (eds), *The Global Financial System*. Cambridge, MA: Harvard Business School Press. (p. 3–31). [10]

Miller, K.D. (1992). A framework for integrated risk management in international-business. *Journal of International Business Studies*, 23, 311–331. [8]

Minsky, H.P. (1982). *Can 'It' Happen Again?* Armonk, NY: M.E. Sharpe, Inc. [4] [5]

Minsky, H.P. (1986). *Stabilizing an Unstable Economy*. New Haven, CT: Yale University Press. [2nd Edition published by McGraw-Hill, 2008.] [4] [5]

Mirowski, P. (1989). *More Heat than Light*. New York: Cambridge University Press. [4]

Mishkin, F. (2009). Not all bubbles present a risk to the economy. *Financial Times* (10 November), p. 11. [9]

Mitroff, I. (2004). *Crisis Leadership: Planning for the Unthinkable*. Wiley. [10]

Monks, R. and Minnow, N. (1991). *Power and Accountability*. New York: Harper Collins. [8]

Morris, C.R. (2008). *The Two Trillion Dollar Meltdown* (revised and updated). New York: Public Affairs. [5]

Nelson, D.B. (1991). Conditional heteroskedasticity in asset returns: a new approach. *Econometria*, 59, 347–370. [7]

Nelson, R.R. and Winter, S.G. (1982). *An Evolutionary Theory of Economic Change*. Cambridge, MA: Harvard University Press. [8]

Newman, M.E.J. (2005). Power Laws, Pareto Distributions and Zipf's Law. *Contemporary Physics*, 46, 323–351. [4] [5] [7]

Nicolis, G. and Prigogine, I. (1989). *Exploring Complexity: An Introduction*. New York: Freeman. [5]

Nietzsche, F. (2007; orig. 1895). *Twilight of the Idols*. Ware, Hertfordshire: Wordsworth Editions Limited. [Prologue].

Odling-Smee, F.J., Laland, K.N. and Feldman, M.W. (2003). *Niche Construction*. Princeton, NJ: Princeton University Press. [5]

Pareto, V. (1897). *Cours d'Economie Politique*. Paris: Rouge and Cie. [4]

Peitgen, H.O., Jürgens, H. and Saupe, D. (1992). *Chaos and Fractals: New Frontiers of Science*. New York: Springer-Verlag. [2nd Edition published in 2004]. [7]

Perrow, C. (1967). Framework for comparative analysis of organizations. *American Sociological Review* 32, 194–208. [8]

Perrow, C. (1984). *Normal Accidents: Living with High Risk Technology*. Princeton, NJ: Princeton University Press. [Pt I]

Perrow, C. (1999). Organizing to reduce the vulnerabilities of complexity. *Journal of Contingencies and Crisis Management*, 7 (3), 150–155. [10]

Peters, E.E. (1994). *Fractal Market Analysis: Applying Chaos Theory to Investment and Economics*. New York: Wiley. [7]

Philippon, T., Reshef, A. (2009). Wages and human capital in the US financial industry 1909–2006. NBER Working Paper Series, Working Paper 14644, National Bureau of Economic Research, Cambridge MA. [8]

Phillips, K. (2008). *Bad Money*. New York: Viking, Penguin Group. [5]

Plerou, V., Gopikrishnan, P. and Stanley, H.E. (2003). Two-phase behaviour of financial markets. *Nature*, 421, 130. [7]

Podobnik, B., Fu, D., Jagric, T., Grosse, I. and Stanley, H.E. (2006). Fractionally integrated process for transition economics. *Physica A*, 362, 465–70. [4]

Poincaré, H. (1914). *Science and Method* (trans. Francis Maitland). London: T. Nelson. [Originally *Science et Méthode*, 1908]. [4] [9]

Poon, S.-H. and Granger, C.W.J. (2003). Forecasting volatility in financial markets: a review. *Journal of Economic Literature*, 41, 478–539. [7]

Popper, K.R. (1967). *The Logic of Scientific Discovery*. 2nd Edition. London, UK: Hutchinson – Radius Books. [2]

Power, M. (2004). *The Risk Management of Everything: Rethinking the Politics of Uncertainty*. London: Demos. [8]

Power, M. (2008). *Organized Uncertainty – Designing a World of Risk Management*. Oxford: Oxford University Press. [8]

Prigogine, I. (1955). *An Introduction to Thermodynamics of Irreversible Processes*. Springfield, IL: Thomas. [5]

Record, N. (2010). How to make the bankers share the losses. *Financial Times*, (6 January), http://www.ft.com/cms/s/0/dda17cc4-fafa-11de-94d8-00144feab49a.html (accessed 15 February 2010). [9]

Redelico, F.O., Proto, A.N. and Ausloos, M. (2008). Power law for the duration of recession and prosperity in Latin American countries. *Physica A*, 387, 6330–6336. [4]

Reinhart, C.M. and Rogoff, K. (2009). *This Time is Different: Eight Centuries of Financial Folly*. Princeton University Press. [6]

Renn, O. (2008). *Risk Governance – Coping with Uncertainty in a Complex World*. London: Earthscan. [8]

Richardson, M. (2009). Causes of the financial crisis of 2007–2009. In V.V. Acharaya and M. Richardson (2009), *Restoring Financial Stability. How to Repair a Failed System*. New York, USA: John Wiley and Sons. [3]

Roberts, J. (2001). Trust and control in Anglo-American systems of corporate governance: the individualizing and socializing effects of processes of accountability. *Human Relations*, 54, 1547–1572. [8]

Roberts, K.H. (1990). Some characteristics of one type of high-reliability organization. *Organization Science*, 1, 60–176. [10]

Rook, L. (2006). An economic psychological approach to herd behavior. *Journal of Economic Issues*, 40, 75–95. [7]

Rosser, J.B. (1994). Dynamics of emergent urban hierarchy. *Chaos, Solitons and Fractals*, 4, 553–562. [4]

Rosser, J.B. (2008). Econophysics. In L.E. Blume and S.N. Durlauf (eds), *The New Palgrave Dictionary of Economics*. 2nd Edition. New York: Palgrave Macmillan. [4]

Rostowski, J. (2010). Intolerance of small crises led to this big one. *Financial Times*, (14 January), http://www.ft.com/cms/s/0/602fd6ee-0079-11df-b50b-00144feabdc0.html (accessed 15 February 2010). [9]

Saichev, A., Malevergne, Y. and Sornette, D. (2009). Theory of Zipf's Law and Beyond. Lecture Notes in Economics and Mathematical Systems, Volume 632. Springer Verlag. [6]

Samuelson, P.A. (1947). *Foundations of Economic Analysis*. Cambridge, MA: Harvard University Press. [4]

Sanderson, R. and Hughes, J. (2009). Banks face change to loan losses rule. *Financial Times* (6 November), 15. [9]

Satinover, J.B. (2002). *The Quantum Brain*. New York: Wiley. [6]

Satinover, J.B. and Sornette, D. (2007). Illusion of control in minority and parrondo games. *European Physical Journal*, B60, 369–384. [6]

Scheinkman, J.A. and Woodford, M. (1994). Self-organized criticality and economic fluctuations. *American Economic Review*, 84 (2), 417–421. [6]

Schmidt, R.H. and Tyrell, M. (2004). What constitutes a financial system in general and the German financial system in particular? In J.P. Krahnen and R.H. Schmidt (eds), *The German Financial System*. Oxford: Oxford University Press. [10]

Schroeder, M. (1991). *Fractals, Chaos, Power Laws*. New York: Freeman and Co. [4] [7]

Sethi, R. (1996). Endogenous regime switching in speculative markets. *Structural Change and Economic Dynamics*, 7, 99–118. [7]

Shackle, G.L.S. (1949). *Expectations in Economics*. Cambridge: Cambridge University Press. [7]

Shapiro, S.P. (1987). The social control of impersonal trust. *American Journal of Sociology*, 93, 623–658. [8]

Sharpe, W.F. (1964). Capital asset prices: a theory of market equilibrium under conditions of risk. *Journal of Finance*, 19, 425–442. [4]

Shiller, R.J. (2008). *The Subprime Solution. How Today's Global Financial Crisis Happened and What to Do about it*. Princeton, USA: Princeton University Press. [3]

Shnerb, N.M., Louzoun, Y., Bettelheim, E. and Solomon, S. (2000). The importance of being discrete: life always wins on the surface. *Proceedings of the National Academy of Sciences*, 97 (9), 10322–10324. [6]

Siegel, J. and Schwartz, J. (2006). Long-term returns on the original S&P 500 companies. *Financial Analysts Journal*, 62 (1), 18–31. [10]

Smith, A. (1999; orig. 1776). *The Wealth of Nations, Books IV-V*. London, UK: Penguin Classics. [3] [4]

Smith, W.K. and Tushman, M.L. (2005). Managing strategic contradictions: a top management model for managing innovation streams. *Organizational Science*, 16, 522–536. [8]

Song, D.-M., Jiang, Z.-Q. and Zhou, W.-X. (2009). Statistical properties of world investment networks. *Physica A*, 388, 2450–2460. [4]

Sornette, D. (1998). Discrete scale invariance and complex dimensions. *Physics Reports*, 297 (5), 239–270. [6]

Sornette, D. (2002). Predictability of catastrophic events: material rupture, earthquakes, turbulence, financial crashes and human birth. *Proceedings of the National Academy of Sciences USA*, V99 SUPP1, 2522–2529. [6]

Sornette, D. (2003a). Critical market crashes. *Physics Reports*, 378 (11), 1–98, Elsevier Science. [6] [7]

Sornette, D. (2003b). *Why Stock Markets Crash?* Princeton, NJ: Princeton University Press. [7]

Sornette, D. (2004). *Critical Phenomena in Natural Science: Chaos, Fractals, Self-organization and Disorder: Concepts and Tools*. Berlin: Springer-Verlag. [7]

Sornette, D. (2006). *Critical Phenomena in Natural Sciences, Chaos, Fractals, Self-organization and Disorder: Concepts and Tools*. 2nd Edition (2nd printing). Springer Series in Synergetics. Heidelberg. [6]

Sornette, D. (2009). Dragon-Kings, Black Swans and the prediction of crises. *International Journal of Terraspace Science and Engineering*, 2 (1), 1–17. [6]

Sornette, D. and Johansen, A. (2001). Significance of log-periodic precursors to financial crashes. *Quantitative Finance*, 1, 452–471. [4] [7]

Sornette, D. and Woodard, R. (2010). Financial bubbles, real estate bubbles, derivative bubbles, and the financial and economic crisis. In M. Takayasu, T. Watanabe and H. Takayasu (eds), *Proceedings of APFA7 (Applications of Physics in Financial Analysis), New Approaches to the Analysis of Large-Scale Business and Economic Data*. Springer. [6]

Sornette, D., Johansen, A. and Bouchaud, J.-P. (1996). Stock market crashes, precursors and replicas. *Journal Physics I France*, 6, 167–175. [7]

Sornette, D., Malevergne, Y. and Muzy, J.-F. (2002). Volatility Fingerprints of Large Shocks: Endogenous versus Exogenous. Available at: http://arXiv:cond-mat/0204626v1 [cond-mat.stat-mech] (accessed 22 October 2009). [9]

Sornette, D., Woodard, R., Fedorovsky, M., Riemann, S., Woodard, H. and Zhou, W.-X. (2009). (The Financial Crisis Observatory), The Financial Bubble Experiment: advanced diagnostics and forecasts of bubble terminations. Available at: http://arxiv.org/abs/0911.0454. [6]

Soros, G. (2008). *The New Paradigm for Financial Markets*. New York: Public Affairs. [5]

Spanos, A. (1986). *Statistical Foundations of Econometric Modelling*. Cambridge, UK: Cambridge University Press. [4]

Stanley, M.H.R., Amaral, L.A.N., Buldyrev, S.V., Havlin, S., Leschhorn, H., Maass, P., Salinger, M.A. and Stanley, H.E. (1996). Scaling behaviour in the growth of companies. *Nature*, 379, 804–806. [4]

Sterman, J. D. (2000). *Business Dynamics. Systems Thinking and Modeling for a Complex World*. New York, USA: McGraw-Hill. [3]

Struzik, Z. (2001). Wavelet Methods in (Financial) Time-series processing. *Physica A*, 296: 307–319. [7]

Suchman, L.A. (1987). *Plans and Situated Actions: The Problem of Human-Machine Communications*. Cambridge University Press, Cambridge, UK. [8]

Summers, L. (1991). The scientific illusion in empirical macroeconomics. *Scandinavian Journal of Econometrics*, 93, 129–148). [4]

Sundström, G.A. and Hollnagel, E. (2006). Learning how to create resilience in business systems. In E. Hollnagel, D.D. Woods and N. Leveson (eds), *Resilience Engineering. Concepts and Precepts*. Aldershot, UK: Ashgate. [10]

Sundström, G.A. and Hollnagel, E. (2011). The importance of functional interdependencies in financial services systems. In E. Hollnagel, J. Pariès, D.D. Woods and J. Wreathall (eds) *Resilience Engineering in Practice*. Aldershot, UK: Ashgate. [10]

Svetlova, E. (2008). Framing complexity in financial markets – an example of portfolio management. *Science, Technology and Innovation Studies*, 4, 115–130. [8]

Taleb, N.N. (2007). *The Black Swan: The Impact of the Highly Improbable*. New York: Random House. [2] [6] [10]

Task, A. (2010). Kill Wall Street Bonuses or Tax 'em to Death, MIT's Simon Johnson Says. *Yahoo! Finance* (12 January), http://finance.yahoo.com/tech-ticker/kill-wall-street-bonuses-or-tax-%27em-to-death-mit%27s-simon-johnson-says-402210.html?tickers=XLF,JPM,GS,BAC,C,MS,WFC (accessed 15 February 2010). [9]

Taylor, F.W. (1911). *The Principles of Scientific Management*. New York: Harper and Row. [8]

Taylor, J.B. (2009). *Getting Off Track. How Government Actions and Interventions Caused, Prolonged, and Worsened the Financial Crisis*. Stanford, USA: Hoover Institution Press. [1] [3]

Tett, G. and Gangahar, A. (2007). System error: why computer models proved unequal to market turmoil. *Financial Times* (15 August), 7. [8]

The Economist (2008). Confessions of a risk manager, vol. 388 (8 August), 72–74[6]

The Economist (2009a). Efficiency and beyond, vol. 392 (8640; 18 July), 71–72. [4]

The Economist (2009b). Too big to bail out, vol. 393 (8654; 24 October), 68. [9]

The Economist (2009c). The other-worldly philosophers, vol. 392 (8640; 26 November), 68–70. [4]

The Economist (2009d). Systems failure, vol. 393 (8659; 26 November), 88. [5]

The Economist (2010). A special report on financial risk, vol. 394 (8669; 13 February): 18 pages after p. 52. [5] [9]

Thompson, J.D. (1967). *Organizations in Action*. New York: McGraw-Hill. [8]

Torrence, C. and Compo, G.P. (1998). A practical guide to Wavelet analysis. *Bulletin of the American Meteorological Society*, 79, 61–78. [7]

Trist, E. (1981). The evolution of socio-technical systems as a conceptual framework and as an action research program. In A. Van de Ven and Joyce (eds), *Perspectives on Organization Design and Behavior*. New York: Wiley, 1981, 19–75. [3]

Turner, J.S. (2000). *The Extended Organism*. Cambridge, MA: Harvard University Press. [5]

Tushman, M.L. and O'Reilly, C.A. (1996). Ambidextrous organizations: managing evolutionary and revolutionary change. *California Management Review*, 38, 8–29. [8]

US reasury (2008). Major Foreign Holdings. Available at: http://74.125.155.132/search?q=cache:NK6nT7agdiEJ:www.ustreas.gov/tic/mfhhis01.txt+http://www.treas.gov/tic/mfh.txt&cd=2&hl=en&ct=clnk&gl=us [5]

Van De Ven, A.H., Delbecq, A.L., Koenig, R. (1976). Determinants of coordination modes within organizations. *American Sociological Review*, 41, 322–338. [8]

Varkoulas, J.T. and Baum, C.F. (1996). Long-term dependence in stock returns. *Economics Letters*, 53, 253–259. [4]

Vasconcelos, G.L. (2004). A Guided walk down Wall Street: an introduction to econophysics. *Brazilian Journal of Physics*, 34, 1039–1065. [4]

Von Bertalanffy, L. (1975). *Perspectives on General Systems Theory*. New York, USA: George Braziller. [2]

Wall, T.D., Cordery, J.L. and Clegg, C.W (2002). Empowerment, performance, and operational uncertainty: a theoretical integration. *Applied Psychology – International Review*, 51, 146–169. [8]

Wasden, C.L. 2010. Taxonomy of Social Tensions Derived from the Global Financial Crisis: An Exploratory Sequential Partial Mixed methods Study. PhD Dissertation, Graduate School of Education and Human Development, George Washington University, Washington, DC. [9]

Watts, D. (2003). *Six Degrees: The Science of a Connected Age*. New York: Norton. [5]

Weber, M. (1947). *The Theory of Social and Economic Organisation*. New York: Oxford University Press. [8]

Weick, K.E. (1976). Educational organizations as loosely coupled systems. *Administrative Science Quarterly*, 21–19. [8]

Weick, K.E. (1995). *Sensemaking in Organizations*. Thousand Oaks, CA: Sage. [8]

Weick, K.E. and Sutcliffe, K.M. (2005). Organizing and the process of sensemaking. *Organization Science*, 16 (4), pp. 409–421. [4]

Weick, K.E., Sutcliffe, K.M. and Obstfeld, D. (1999). Organizing for high reliability: processes of collective mindfulness. *Research in Organizational Behaviour*, 21, 81–123. [8]

Weisstein, Eric W. (no date). Zöllner's Illusion. From *MathWorld* – A Wolfram WebResource. Available at: http://mathworld.wolfram.com/ZoellnersIllusion.html (accessed 29 December 2010). [4]

Weitz, E. and Shenhav, Y. (2000). A longitudinal analysis of technical and organizational uncertainty in management theory. *Organization Studies*, 21, 243–266. [8]

Wellink, Nout. (2010). Fundamentally strengthening the regulatory framework for banks. Bank for International Settlements, September. [10]

West, B.J. and Deering, B. (1995). *The Lure of Modern Science: Fractal Thinking*. Singapore: World Scientific. [4] [5] [7]

Westrum, R. (1993). Cultures with requisite imagination. In J. Wise, V.D. Hopkin and P. Stager (eds), *Verification and Validation of Complex Systems: Human Factors Issues*. Berlin: Springer-Verlag. [3]

Whorf, Benjamin Lee. (1940). Linguistics as an exact science. *Technology Review*, 43, 61–63, 80–83. Reprinted in John B. Carroll (ed.) (1956), *Language, Thought, and Reality*, 220–232. Cambridge, MA: Technology Press. [3]

Wiener, N. (1948). *Cybernetics: Or Control and Communication in the Animal and the Machine*. Cambridge, MA: MIT Press. [2nd revised Edition 1961.] [3]

Wolfson, M.H. (2002). Minsky's theory of financial crises in a global context. *Journal of Economic Issues*, 36, 393–400. [5]

Wong, J.C., Lian, H. and Cheong, S.A. (2009). Detecting macroeconomic phases in the Dow Jones industrial average time series. *Physica A*, 388, 4635–4645. [4]

www.jchs.harvard.edu/publications/./son2008/index. [4]

Yaari, G., Nowak, A., Rakocy, K. and Solomon, S. (2008). Microscopic study reveals the singular origins of growth. *The European Physical Journal*, B62 (4), 1434–6028. [6]

Yakovenko, V.M. and Rosser, J.B. (2009). Colloquium: statistical mechanics of money, wealth, and income. *Reviews of Modern Physics*, 81, 1703–1725. [4]

Yalamova, R. (2003). Wavelet MRA of Index Patterns around Stock Market Shocks, PhD thesis, Kent State University. [7]

Yalamova, R. (2010). Stock Market Index Dynamics before Crashes: A Time-scale Adaptive Research Framework. Working paper, Faculty of Management, University of Lethbridge, Canada. [7]

Yalamova, R. and McKelvey, B. (2011). Explaining what leads up to stock market crashes: a phase transition model and scalability dynamics. *Journal of Behavioral Finance*, in press. [7] [9]

Yan, W., Woodard, R. and Sornette, D. (2010). Diagnosis and prediction of tipping points in financial markets: crashes and rebounds. *Physics Procedia*, 00,1–17. [4] [7] [9]

Yergin, D. (2009). A crisis in search of a narrative. *Financial Times*, (20 October), http://www.ft.com/cms/s/0/8a82d274-bda9-11de-9f6a-00144feab49a.html (accessed 24 October 2009). [9]

Zanini, M. (2008). Using power curves to assess industry dynamics. *McKinsey Quarterly* (November) 1–6. [4]

Zhang, F. (2006). Information uncertainty and stock returns, *Journal of Finance*, 61, 105–37. [7]

Zhang, J., Chen, Z. and Wang, Y. (2009). Zipf distribution in top Chinese firms and an economic explanation. *Physica A*, 388, 2020–2024. [4]

Zhou, W.-X. and Sornette, D. (2003). Non-parametric analysis of log-periodic precursors to financial crashes. *International Journal of Modern Physics C*, 14, 1107–1126. [4]

Zipf, G.K. (1929). Relative frequency as a determinant of phonetic change. *Harvard Studies in Classical Philology*, 40, 1–95. [4]

Zipf, G.K. (1932). *Selective Studies and the Principle of Relative Frequency in Language*. Cambridge, MA: Harvard University Press. [4]

Zipf, G.K. (1949). *Human Behavior and the Principle of Least Effort*. New York: Hafner. [4]

Zunino, L., Zanin, M., Tabak, B.M., Pérez, D.G. and Rosso, O.A. (2009). Forbidden patterns, permutation entropy and stock market inefficiency. *Physica A*, 388, 2854–2864. [9]

Index

2007 liquidity crisis
  analysis 48–54
  introduction 41–2
  'Minsky Moments' 43–6, 53
  positive feedback 43–6
  scalability dynamics 1972–2007 42–3
  scale-free theories 46–8
  see also Great Recession
2007–2009 financial crisis
  description 1–2
  linear view 2–6
  phases 4–5
  scope and impact 6
  what happened? 2

Adrian, Tobias 82
Against Individual Agents (resilience) 138,
  140–1
AIG 50, 141
airplane cockpit control systems 154
ARCH (stochastic process) 98
autocorrelation function and power law
  86, 96–100
avalanches
  power law 74
  probability density function 75

Bachelier, Louis 27–8, 43, 133, 147–8
Bank of England 1, 142
Bank for International Settlements, Basle
  81
Bank of Japan 1
Barney Smith 50
Basel II accord 122, 151
Basel III accord 150–1, 163
Bear Sterns 53, 82, 148
Bernanke, Ben 81, 136, 140, 143
'Black Monday' 41
Black Swan logic
  description 12–13
  Dragon Kings 68–9, 71

Black-Scholes options-pricing model 49,
  54, 133
Brady, Nicholas 148
Bretton Woods agreement 125
'Brownian motion' (volatility) 99
bubbles
  build-up
    financial markets 87, 95
    Hurst exponent 147
    interventions 29, 133
    positive feedback 45
    resilience interventions 108, 133,
      138, 140
    tiny initiating events 54,
      volatility 99, 134
  couplings 24–5
  financial experiment 79–80
  force 145
  housing 24, 51, 53
  Hurst exponent 147
  prediction 39, 56
  prevention 81
  reckless endangerment 140
  stock market 35, 56, 85–105
  tiny initiating events 54
Burger King 71

Capital-Asset-Pricing Model (CAPM) 29
casino tax *see Tobin tax*
caulilflower
  fractal structure 34, 99–100
  Square-Cube Law 34
CDOs *see* collaterised debt obligations
Citigroup 50
Classical Financial Theory 104
Clinton administration 50
closed systems 14–16
CMOs *see* collaterised mortgage
  obligations
cooperativity in markets 62–5
collaterised debt obligations (CDOs) 3,
  141

collaterised mortgage obligations
      (CMOs) 3
combination theory 42, 48, 50, 52, 54
complex systems 58–9, 96, 154
'computer herding' 124
confidence of valuations 23–4
connectivity (positive feedback and
      Minsky Moments) 45–6
contagion bursts 47, 49, 50–3
contagion of FE formulas 143
control
      decision-making 117
      description 151
      environments 154
      feedback loops 14–15, 20
      illusion 65–6
      limits 131–2
      theory 19
correlated behaviours 29
coupled systems
      bubbles 24–5
      Financial Services system 16
      'invisible hand' 22–3
      processes 21–2
      second cybernetics 26
Cox, Christopher 145
crashes *see* market crashes
Critical Point
      Dow Jones index 93
      financial market crashes 96, 99, 104
      information and noise 101
      free markets 136
      market crashes 134, 136
      relaxation time 135
      resilience dynamics 148
      Triple Point 90–2
cybernetics 17–19, 26
      see also second cybernetics

Debt Obligations (DOs) 60
decision-making
      control 117
      independent/rational 101, 104
      individual traders 125
      information complexity 86
      investment banking 112
      noise in markets 88, 91
      nuclear power plants 112
      option evaluation 38
      process rules 129
      responsible 132

social interaction 124
      uncertainty 121
demand (economics) 18
derivatives 41–2
'deviation-amplifying mutual causal
      processes' 21, 26
Don Quixote 55
DOs *see* Debt Obligations
'dot.com' bust 41
Dow Jones index (DJI)
      autocorrelation function 98
      fractal region 93
      power law 34, 92, 96
      short selling 143
Dragon Kings
      Black Swans 68–9, 71
      crashes 76–8
      emergence 71
      Financial Services system 157
      identification 39
      Paris 69–71
      positive feedback 70
      power law 72, 76
      spontaneous recovery 84
      taming manias 67–8
driving cars analogy 138, 147
Dudley, William 81–2
dynamic systems modelling
      introduction 17–18
      'invisible hand' 18–19, 22–3
      pro-cyclicality 24–5
      second cybernetics 20–1, 26

early warning signs 109, 149
ECB *see* European Central Bank
economy and negative feedback 20–1
econophysics
      classical economics 31–2
      Financial Services systems 29
      fractal structures 33–4
      Hurst exponent 134
      key elements 28, 30–2
      power law 33–4, 134
      resilience engineering 32–5
      scale-free theories 34–5
efficiency-thoroughness trade-off 12
Efficient Market Hypothesis (EMH)
      'Brownian Motion' 99
      description 29, 87, 91
      market behaviour 87, 97, 33, 134–5
      model 157

normal markets 39, 85
random-walk 148
regulation 147
tipping point 35, 97
Triple Point 93, 135, 138, 157
volatility 99, 135
EGARCH (stochastic process) 98
emergence
   2007–2009 financial crisis 24, 35
   complex structures of finance
      models 96
   Dragon Kings 71
   skewed distributions of stock-
      market volatilities 42
endogenous shocks 135
European Central Bank (ECB) 1, 20, 140
European Systemic Risk Board 1
exogenous shocks 135
Expert systems and credit-worthiness of
   customers 123

Fanny Mae 50, 52, 141
FBE *see* financial bubble experiment
FDIC (Federal Deposit Insurance
   Company) 5, 142
'fear and greed' 135–6
Federal Reserve Bank, US
   Bernanke, Ben 81, 136, 140, 143
   discount rate 52–3, 140
   financial engineering 145
   Greenspan, Alan 19
   low interest policy 24
   reckless endangerment 140
   regulatory bodies 1
   rescue plan 41
feedback-control loops 14–15, 20
financial bubble experiment (FBE) 79–80
financial crisis *see* 2007–2009 financial
   crisis
financial engineering (FE)
   computer-based 41
   formulas 143, 147
   methods 133
   resilience 137
financial markets
   crashes, 96, 99, 104, 135–6, 143, 147
   Critical Point 90–2, 96, 99, 104
   Hurst exponent 25, 32, 39, 99–100,
      103, 105, 147
   information and noise 101
   Law of Requisite Variety 17–18, 165–8

phase diagram 87–89, 89–90, 99
regions (phase diagram)
   certainty 92
   fractal 93
   risk 93
   uncertainty 92–3
scale-free perspective 86–94
tipping point 101–5
Triple Point
   Critical Point 90–2, 97, 99
   dynamics 89–90
   financial markets 87–8, 102
Financial Services Authority, UK 149
Financial Services system (FSS)
   analysis 9–10
   components 15
   control systems 154
   coupling 16
   crisis dynamics 23
   definition 14
   description 11–16
   Dragon Kings 157
   functional perspective 12–13
   goal 13
   institutional perspective 12
   pro-cyclicality 24
   regulation, governance and control
      conclusions 155–62
      description 150–4
      introduction 149–50
      regulatory bodies 1–2
      resilience engineering 155–62,
         162–5
   resilience engineering 8–10, 155–65
   resilient systems 107–9
   self-regulation 14
   tight couplings 107
   tractability 9, 11, 13–14
Financial Stability Board (FSB), US 24
Financial Stability Oversight Council
   (FSOC), US 151–2
Financial Times 139
First Axiom of Industrial Safety 2
foreign exchange trading and managing
   uncertainty 125–7
formulaic automaticity 137
*formulaic resilience engineering
   interventions* 140, 144–6
Forrester, Jay W. 21
Foundations of Economic Analysis 30
fractal structures

econophysics 33–4
financial markets 93
Hurst exponent 99–100
power law
    autocorrelation function 96–8
    distribution of returns 95–6
resilience engineering 167
see also cauliflower
FRAM (Functional Analysis Reference
        Method) model for FSS 159–61, 163
Freddie Mac 50, 52, 141
FSOC *see* Financial Stability Oversight
        Council
FSS *see* Financial Services system
functional perspective 157–8
functional perspective (Financial
        Services system) 12–13

G20 Finance Ministers' meeting, St.
        Andrews 142
Galileo 34
GARCH (Generalised Autoregressive
        Conditional Heteroskedasticity)
        93–4, 98
Gargantua and Pantagruel 28
General Systems Theory 14, 18–19
Gibrat's Law 69–70
Glass-Act, US 42, 148
Goldman Sachs 50, 81, 139
governance objective 151
gravity and financial crises 45
Great Depression 41
'*Great Moderation*' period 28
Great Recession (2007 liquidity crisis)
    causes 41–2, 136
    '*correlated behaviours*' 29
    explanation 135–6
    financial losses/unemployment 94
    market crashes 135
    scalability spirals 133
    timeline 53–4
Greek Sovereign debt 139
Greenspan, Alan 19, 29
'Greenspan put' 144

Hang-Seng stock market 103
'herding' behaviour
    autocorrelation function 97, 99
    bubble build-up 140
    crashes 28
    decision-making 86

econophysics 133
Efficient Market Hypothesis 147
initiating events 45, 133
interventions 29
financial markets 134
noise trading 88
positive feedback 77
price-setting mechanism 76
resilience via information disclosure
        138, 140
scalability dynamics 54
tipping point 35, 91, 101, 104–5
unhealthy market 39
volatility 99
homo economicus markets 7, 38
housing bubbles 51
Hurst exponent
    bubble build-up and market crashes
        147
    econophysics 32, 134
    financial markets marker 25, 32, 39
    Financial Times 139
    formulaic resilience interventions 144
    fractality 99–100
    herding behaviour 140
    Nile River dam size 100
    Tipping Point 103, 105, 138
    volatility 144
    'volatility autocorrelation function' 35

IBM PC 49
IGARCH (stochastic process) 98
illusion of control 65–6
independent and identically distributed
        (*i.i.d.*) behaviour 28, 43
information complexity threshold and
        tipping point 101–5
Information Disclosure (resilience
        interventions) 138–40, 146
institutional perspective 157–8
institutional perspective (Financial
        Services system) 11
Insurance (resilience interventions) 138,
        141–3, 144, 147
International Accounting Standards
        Board 139
International Bank for Settlements 150
International Monetary Fund (IMF) 142
intractability (management and
        governance) 107
investment banking (decision making) 112

'invisible hand'
    description 18–19, 21
    visibility 22–3
irregularity generated gradients 47–8, 49

Johnson, Simon 148

Key Performance Indicators (KPI) 150,
    153, 161
King, Mervyn 142

Law of Requisite Variety
    dynamic systems modelling 17–18
    financial markets 165–8
Lehman Brothers 82, 147–8
leverage 82–3, 142
Lévy distributions 31
Lewis, Ken 143
LIBOR (banking rates) 4
linear view of financial crisis 2–6
linguistic realism theory 17
log-periodic (nonlinear oscilllatory)
        power law (LPPL) 28, 78–9, 92, 99
Long-term Capital Management
        (LTCM) 27, 41, 49, 62

M&M Peanuts (sandpile analogy) 45
Macdonald's 71
Madoff Ponzi scheme 145
market crashes
    Critical Point 99, 134
    detection/prediction 78–80
    Dragon Kings 76–8
    see also financial markets
matter (solid, liquid and gas) 86
MBSs *see* mortgage-backed securities
Merrill Lynch 4, 143
Merton, Robert 27
mini-Goldmans 148
Minority Game (MG) 62–3, 63–5, 66, 77
'Minsky Moments' 43–6, 53
Minsky, positive feedback processes 54
Mirowski, P. 30
monitoring (resilience engineering and
        FSS) 154, 155–7
More Heat Than Light 30
mortgage market, US 3
mortgage-backed securities (MBSs)
    Dragon Kings 76
    financial instruments 41–2
    invention 50

'junior tranche' 61
loan selling 3
Minority Game 65
scale-free theories 47
security tranches 82
tiny initiating events 52
toxicity 53
unanticipated coupling 60
mutually coupled processes 22–3

NASA accidents 121
NASDAQ 76, 78, 92, 102
negative feedback 20–1
New Century Financial Corporation 149
New York Federal Reserve Bank 81–2
Nile River dam size 100
noise and information (financial
        markets) 101–2
Noise to Information trading ratio 88
Northern Rock 53
nuclear power plants (decision-making)
        112

Obama, President 142
open systems 14–15, 17
Other interventions (resilience) 138, 144
out of equilibrium system 72

Pareto distribution 31, 33, 50, 54, 100
Paris: Dragon King of French cities
        69–71, 76
Paulson, Henry 54
People's Bank of China 1
performance variability (FRAM) 159
phase diagram (financial markets)
    Critical Point 94, 99
    regions 92–3
    Triple Point 87–89, 94, 99
    Triple Point dynamics 89–90
phase transition
    2007 liquidity crisis 49, 51
    model of market dynamics 85–6
    Triple Point 87, 102
Physica 104
Poincaré, Henri 27, 134
Poland (economic growth) 84
positive feedback
    2007–2009 financial crisis 39
    Dragon Kings 70
    leverage 82
    'Minsky Moments' 43–6, 54

second cybernetics 20–1
power law
    autocorrelation function 86, 96–100
    avalanches 74
    distribution of returns 95–6
    Dow Jones Index 34, 92, 96
    Dragon Kings 72, 76
    econophysics 28, 33–4, 41, 54, 134
    fractality and Hurst exponent 100
    log-periodic (nonlinear oscilllatory)
      78–9
    market capitalisation 34
    resilience engineering 167
    self-organisation 34
    stock price volatility 85, 85–6
    tension and connectivity 46
    Zipf 34, 70
prediction of bubbles 39, 46
preferential attachment 48, 52–3
pro-cyclicality (dynamic systems
    modelling) 24–5
probability density function 75

quantitative models for trading
    decisions (uncertainty
    management) 123–4

'random directed polymer' (RDP) 73–4
'rank/frequency' effect 33
reckless endangerment
    financial services firms 35
    resilience interventions 138, 140–1,
      144, 147
regulation
    bubble-prediction 56
    definition 150–1
    Financial Services system 14, 153–4
    resilience engineering 169
    Triple Point 147
    see also self-regulation
relaxation time 134–5
requisite imagination 18
Resilience Analysis Grid 164
resilience creation
    against individual agents 140–1
    information disclosure 138–140
    insurance 141–3
    returning to normal 144–6
    systemic connectivities 143–4
resilience engineering
    assumptions 13

description 8–10
'deviation-amplifying mutual causal
    processes' 26
Dragon Kings 67
econophysics 32–5
Financial Services system
    conclusions 162–4
    function-based representation 159
    functional model 159–61
    institutional/functional
      perspectives 157–9
    monitoring 155–7
    responding 162
past events 78
process dynamics 167
risk/reckless endangerment
    individual agents 140–1
    information disclosure 138–40, 146
    insurance 141–3, 144
    internal versus external shocks
      134–6
    introduction 133–4
    other interventions 144
    resilience creation 136–46
    returning to normal 144–6
    summary 146–8
    systemic connectivities 143–4
    traders and bankers 140–1
stability and flexibility in
    organisations 117
summary 167–8
resilience engineering formulas 147
resilience interventions
    concept 108
    Individual Agents
      Insurance 138
      Reckless Endangerment 138
    Information Disclosure 138
    Systemic Interconnectivities 138
risk
    diversification 58–9
    management 113–14
    systemic 1, 149, 152, 154
    uncertainty distinction 113
Roll, Richard 50
root cause perspective and financial
    crises 2

Samuelson, P.A. 30
sandpile analogy (financial crises) 45–6,
    77, 136, 157

Sante Fe vision 32, 34
scalability dynamics 1972Ä2007 42–3
scale-free theories (SFTs)
    2007 crisis 46, 48–50
    combination theory 42, 48, 50, 52, 54
    connectivities 42
    contagion bursts 47, 49, 50–3
    econophysics 33–5, 38–9
    irregularity generated gradients_ 49
    'Minsky Moments' 53
    phase transition 49, 51
    preferential attachment 48, 52–3
    scalability dynamics 43
    spontaneous order creation 48, 50–2
Scholes, Myron 27–8
second cybernetics (dynamic systems
    modelling) 20–1, 26
Securities and Exchange Commission
    (SEC), US 141, 145, 162
securitisation repackaging 143
self-organised criticality
    Dragon Kings 71–2
    sandpiles 45
self-regulation and Financial Services
    system 14
sense-making
    decision process 121
    organisations and complex systems 38
'shadow banking system' 11–12
shadow banking system 148
shadow operations 143
shadow trading 143–4
Shanghai Stock Index 78
sharp irregularity and financial crises 45
Shin, Hyun 82
short selling 142–3
Smith, Adam 18–19, 30
'*social contagion*' of boom thinking 25
spontaneous order creation 48, 50–2
spontaneous recovery 82–3
Square-Cube law 34
stability creates instability 43, 53
stock market bubbles 35, 56, 85–105
subprime market, US 2–4
supply (economics) 18
systemic instabilities 82
Systemic Interconnectivities
    contagion of FE formulas 143
    international changes 144
    resilience interventions 138, 147
    securitisation repackaging 143

shadow trading 143–4
systemic risk 1, 149, 152, 154

taming manias
    Black Swans and Dragon Kings 68–9
    cooperativeness in markets 62–5
    crashes: Dragon Kings of the market
        76–8
    detection and prediction of crashes
        in markets 78–80
    Dragon Kings 67–8
    illusion of control 65–6
    introduction 55–8
    Paris: Dragon King of French cities
        69–71
    policy making in aftermath 80–2
    risk diversification 58–9
    self-organised criticality and Dragon
        Kings 71–2
    spontaneous recovery 82–3
    unanticipated coupling 60–2
teaser loans 48, 52–3, 139–40, 140, 146
tension (positive feedback and Minsky
    Moments) 45–6
The Economist
    capital ratios 142
    'Confessions of a Risk Manager' 60
    efficient market/random walk
        hypotheses 43
    financial risk 147
    Long-term Capital Management 27
    market crashes 28
The Guardian 19
The Wealth of Nations 30
thermostat control (closed systems) 20
TIEs *see* 'tiny initiating events'
tight couplings
    complex systems 58
    Financial Service system 107
tiny initiating events (TIEs)
    2007–2009 financial crisis 38–9
    bubbles 54
    computers and programming 49
    'deposit-style' banks 50
    herding behaviour 133
    houses
        price 51
        value 50
    mortgage defaults 53
    'spiralling causal dynamics' 39, 42, 46
    subprime mortgages 52

tipping point
  Efficient Market Hypothesis 35, 97
  identification 28, 41, 134
  information complexity threshold
    101–5
  R1 102, 137–9, 141–2, 144–8
  resilience engineering 137
  resilience interventions 108
  shadow operations 143
  shadow trading 144
  trading volatilities 140
Tobin, James 137
Tobin tax 137, 142, 146
too big to fail principle 123, 148
tractability
  Financial Services system 9, 11, 13–14
  mathematical 61
trader states (wait, buy and sell) 86–7
Triple Point
  Critical Point 90–2, 99
  definition 85
  Efficient Market Hypothesis 93, 135,
    138, 157
  'fair' prices 103
  financial markets 87–9, 89–90
  free markets 136
  formulaic resilience interventions 144
  Hurst exponent 104
  information cost 97
  noise in markets 102
  phase transition 87, 102
  regulation 147
  relaxation time 134
  resilience 136–7
  risk, uncertainty and certainty 93
Turner, Lord 137

unanticipated coupling 60–2
uncertainty management (financial
    services)
  Basel II accords 122
  foreign exchange trading 125–7
  incentives for responsible behaviour
    130–1
  introduction 111–14
  minimising/coping 114–15
  organisational governance 128
  quantitative models for trading
    decisions 123–4
  regulation 128–30

stability and flexibility in
    organisations 116–17
  strategy 108, 117–22
  too big to fail principle 123
  trust and limits of control 131–2
unexpected events 16
United Kingdom (UK) Financial
    Services Authority 149
United States (US)
  Clinton administration 50
  FDIC 5, 142
  Financial Stability Board 24
  Financial Stability Oversight Council
    151–22
  firms and power law 34
  Glass-Steagall Act 42, 148
  Madoff Ponzi scheme 145
  mortgage market 3
  Office of Comptroller of the
    Currency 151
  regulation 42
  Securities and Exchange
    Commission 141, 145, 162
  subprime market 2–4
  Systemic Risk Regulator 149
  Systemic Risk Regulatory Council 1
  see also Federal Reserve Bank

'value-at-risk' (VAR) model 27
volatility
  Efficient Market Hypothesis 99, 135
  memory 98
  random 99
  PL 144
  temporal conjunction 96
'volatility autocorrelation function' 35
Volcker Rule 148
von Bertalanffy, Ludwig 14

What-You-Look-For-Is-What-You-Find
    (WYLFIWYF) principle
  description 7
  'making sense' 37–8
  naive realism 17
  summary 166
  what happened? 2

Zipf's power law 34, 70
Zöllner's illusion 37–8